Why Elephants & Fleas Don't Sweat

A Zoological Miscellany

Gideon Louw

Illustrated by Dirk van Wyk

Detselig Enterprises Ltd.

Calgary, Alberta, Canada

Why Elephants and Fleas Don't Sweat

© 1996 Gideon Louw

Canadian Cataloguing in Publication Data

Louw, Gideon, 1930-
    Why elephants and fleas don't sweat

ISBN 1-55059-125-8

    1. Energy metabolism. 2. Animal ecophysiology. 3. Physiol-
ogy, Comparative. I. Title.
QP176.L68 1996   591.1'33   C96-910155-4

Detselig Enterprises Ltd.
210-1220 Kensington Rd. N.W.
Calgary, Alberta   T2N 3P5

Detselig Enterprises Ltd. appreciates the financial support for
our 1996 publishing program from the Department of Canadian
Heritage and the Alberta Foundation for the Arts, a beneficiary of
the Lottery Fund of the Government of Alberta.

Printed in Canada   ISBN 1-55059-125-8   SAN 115-0324

Contents

Preface   v

1. The cost of flying   6

2. The cost of swimming 22

3. Fuels for beating entropy 32

4. Napoleon's water molecule 52

5. The thermostat and why elephants and
    fleas don't sweat 62

6. Enough energy for sex? 84

7. Finding the way 98

8. African safari and the sprinting
    compost heap 112

9. Desert sands 126

10. Humans bend the rules:
    from silicone tool to silicone chip 146

Suggested reading 167

Glossary 169

Preface

This book is not meant for professional biologists, although some may have fun reading it. It is meant for lay-persons of all ages who are interested in natural history books that go beyond just pretty colour pictures. It is especially intended to excite the interest of young students interested in how animals work, or more precisely, what makes them tick when they are at work.

The various chapters have been written around the theme of the energy requirements of animals, or less formally, their cost of living. Because energy use is involved in almost all aspects of animal life, I have unashamedly digressed from the main theme to touch on examples of special interest.

For example, when you read this collection, you will learn, among other things: how some fish keep their brains warm, how wolves keep their brains cool, how desert ants navigate, why rhinoceros' milk is so dilute, why bees are so busy, why fleas and elephants don't sweat and that modern civilization owes its origins to chewing the cud.

I wish to thank Tara Gregg and Linda Berry for helpful editorial suggestions. Special thanks are due to my wife, Claire, who encouraged me to write this volume in the first place and provided help in numerous ways.

Gideon N. Louw, Calgary, 1996.

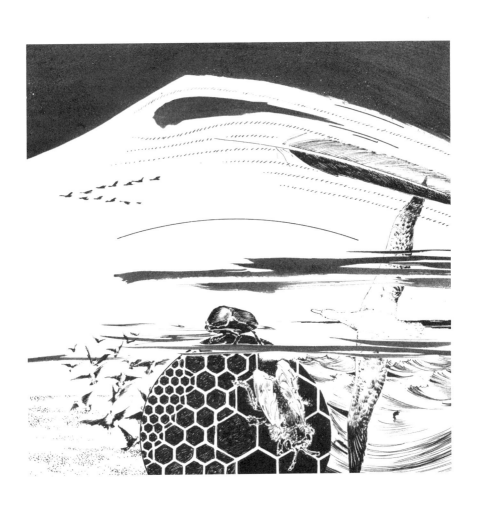

The
Cost
of
Flying

We humans have always been fascinated by the seemingly effortless flight of birds. Perhaps this is because our own mode of locomotion, walking or running, is rather clumsy and demands so much energy. We also envy the birds their panoramic view of the world.

No wonder that we have tried to copy bird flight, from the early wax wings of Icarus in Greek mythology, which melted when he flew too close to the sun, to the sleek mantis-like Concords that thunder across our skies at more than two football fields per second. These modern jet-propelled aircraft are as cost efficient as the larger birds, when efficiency is measured on the basis of body weight transported per unit of distance. In fact, they are much more efficient than most flying insects in this regard, but remain very clumsy when compared to the brilliant aerobatics birds and insects can perform. Let us restrict our discussion to the latter and begin by indulging in some bird watching to appreciate the variety and beauty of some flight patterns.

There are still many excellent localities for bird watching left on our planet. Perhaps one of the best is the lagoon near Walvis Bay on the desert coast of Namibia. Here one is treated to frequent spectacular sunsets over the Atlantic. The persistent fog bank looming over the cold ocean enhances the beauty of the sunset by refracting a splendid array of colours in the red portion of the spectrum from pinks to rose to scarlet and deep red. But the birds using the last few hours of light really fire the imagination. Huge flights of flamingos fly directly overhead. The rush of their wings is truly thrilling and, as the large formation wheels against the westering sun, the dying light brings out the brilliant crimson on the leading edge of their wings. The formation changes shape continuously from the classical V-shape to a boomerang or delta-wing. They then form up line-abreast, and finally line-astern as they disappear to their sleeping grounds.

Early next morning a flotilla of pelicans may sail into your view like a group of sailing ships. They are co-operating by herding small fish together before scooping them up in their absurdly large beaks, seemingly designed for illustrations in children's books. The pelicans will dabble around until the sun has warmed the surrounding desert sufficiently to create rising air masses or thermals. The birds will then leave the water to engage in thermal soaring, drifting and gliding on the rising air until they reach a threshold altitude. This will allow them to glide with a minimal use of energy to some distant fishing ground.

In the distance you will often see large squadrons of Cape Cormorants flying close to the ocean's surface while fanning out over the lagoon, like greedy army ants in a tropical rain forest. They are flying close to the water to exploit the updraft of air on the windward side of the waves.

You might even be lucky enough to enjoy the sight of a greater kestrel attacking an isolated tern. The tern gives a superb display of aerobatics – power dives, flip overs, steep turns and rolls off the top – to finally win the duel. The kestrel then retreats to its more accustomed terrestrial home, using dignified straight and level flight.

All these flying patterns require large amounts of energy when measured as cost per unit time (calories/hour), but when we calculate cost efficiency in terms of units of body weight moved per unit of distance (calories per kilogram per kilometre), flying turns out to be an efficient form of transport. It is far more efficient than walking or running but, as we shall see later, less efficient than swimming.

There are several reasons for this high efficiency. One of the most important is that flying animals, unlike walking and running animals, do not need to interrupt their progress to brace their knees with each forward step, in order to counter gravitational forces. In addition, birds are usually swift flyers and reach their destinations sooner than running animals, thereby using less total energy. Finally, many large birds are able to interrupt high energy flapping flight with long periods of gliding. The latter provides them with lift against gravitational forces and is a most economical form of transport.

During flapping flight the wings of birds are moved both forwards and downwards. In this way air is forced downwards and backwards, creating an upward thrust to provide lift and a forward thrust to propel the bird forwards. Birds engaged in flapping flight are in fact rowing through the air at great energetic cost per unit of time because, unlike aquatic animals, they do not enjoy the buoyancy effect of water. During gliding the wings are not flapped; lift is provided by air moving faster over the curved upper surface of the wing than it does over the flatter surface on the

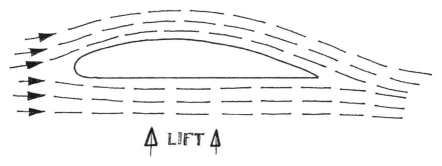

Air moving over the upper surface of a wing moves faster than over the lower surface. Air pressure is therefore reduced on the upper surface, resulting in a positive "lift" force on the lower surface.

underside. The faster moving air on the upper surface has a lower air pressure than the slower moving air on the lower surface; this results in an upward force or lift, which acts on the under surface with minimal effort required from the bird. A gliding bird will, however, lose altitude and eventually have to resort to flapping flight unless it can exploit some form of updraft.

Different species of birds vary markedly in their fuel efficiency. An excellent example of this can be found by comparing two marine birds which look rather similar, the wandering albatross and the Cape gannet. The albatross can stay on the wing for weeks at a time while engaging in what is known as oceanic or dynamic soaring. This means the birds can glide for hours on end, as they exploit the differences in wind speed within the boundary layer of air just above the surface of the ocean, while seeking out their prey.

Within this boundary layer of air, the wind speed is at its lowest immediately above the surface of the ocean, and albatrosses gain speed by gliding downwind and towards the surface of the water. When they are close to the surface, they turn into the wind, which provides them with lift and allows them to gain altitude into the higher layers of faster moving air. In this way they gain altitude while keeping their airspeed high enough to maintain lift on the underside of their wings. This is accomplished without having to resort to flapping flight. When they reach an altitude of about 12 metres above the ocean, they turn downwind once again to repeat the manoeuvre.

In contrast, the Cape gannet forages for fish by using flapping flight to gain sufficient altitude over the ocean, from where it can power-dive upon its prey, swimming just beneath the surface of the ocean. Ornithologists at the University of Cape Town have shown that, because of this energy-demanding lifestyle, the total cost of free existence for the gannet is almost three times that of the albatross.

The birds that use by far the greatest amount of energy for flying are also the smallest, namely hummingbirds, the smallest of which are about the same size as the largest flying moths (2 grams or 0.07 oz). Hummingbirds not only suffer from the disadvantage of requiring large amounts of energy for their characteristic hovering flight while drawing off nectar from flowers, but, because they are so small, they cannot store significant amounts of energy.

Simple geometry tells us that the smaller an object is, the greater its surface area is relative to its volume. Imagine the differences between a football and a marble in this respect. The football has a greater total surface area than the marble, but its surface area relative to its volume or size is much smaller than that of the marble. Therefore, because of their small volume and relatively high surface area, these miniature birds lose heat rapidly at night, and are obliged to enter a state of nocturnal torpor to save energy.

In this condition their metabolism slows down and their body temperatures fall rapidly to take on the temperature of the surrounding air. If the latter, however, falls too low, the metabolic rate of the hummingbirds will increase to arrest the rate of cooling.

One of the most imaginative measurements ever made of hummingbird metabolism was made by Professor 'Bart' Bartholomew, assisted by John Lighton at the University of California in Los Angeles. They modified the standard nectar bird-feeder for hummingbirds by fitting an open mask around the nectar spout. This allowed them to pump the expired air from a hovering hummingbird through the mask and into a system of tubes leading to an oxygen analyzer, connected to a computer. All this took place outside Bart's breakfast room window, from where he has been observing hummingbirds for many years while enjoying his breakfast.

Using oxygen consumption as a measure of energy turnover (or metabolic rate), Bart and John confirmed that hummingbirds pay a very high cost for their hovering flight, and because of this, concluded that the birds seem to walk an energetic tight-rope. This was recently confirmed by Professors Dudley and Peng Chai at the University of Texas, who showed that the ruby-throated hummingbird has a power output of 133 watts per kilogram of muscle compared to a production of only 15 watts per kilogram by human muscle.

Consequently, it has puzzled ornithologists for a long time that hummingbirds do not spend more time foraging for nectar and less time perched on a suitable branch, apparently goofing off. Jared Diamond and his co-workers, also at the University of California in Los Angeles, recently discovered the answer to this puzzle. They fed hummingbirds a cocktail of radioactive isotopes, which allowed them to measure the time it took for a volume of nectar to pass through the stomachs of these small creatures. Their results showed clearly that hummingbirds are forced to rest between foraging forays to allow their stomachs to first empty, before they forage for more nectar.

Most of the nectar consists of water, which contains no food energy, but this inert bulk must also be processed before the birds can take their next meal. It follows then that the sugar concentration of the nectar is critical. If it falls beneath a certain threshold, the birds will spend most of their time collecting and processing water through their digestive tracts. One can assume, however, that the plants providing the nectar have evolved through natural selection to produce nectar with sufficient sugar, otherwise they would not benefit from pollination by the hummingbirds as they move from flower to flower. This is an example of ancient mutual dependence.

The bumblebees in your garden enjoy a similar relationship with their host plants. They also have to roost frequently to first process their watery nectar before proceeding with the next foraging bout. Bumblebee foraging has been intensively studied by Bernd Heinrich of Vermont University and I recommend his informative and readable text, *Bumblebee Economics*.

Professor Bertsch of Marburg University in Germany, another renowned expert on bumblebees, has built an artificial meadow in his laboratory to study foraging bumblebees.

It consists of a large climate chamber, in which light, temperature and humidity are controlled. The walls are covered with bright wallpaper depicting beautiful alpine meadows, while artificial flowers, fitted with miniature nectar pumps, have been placed at strategic intervals on the floor. The entire set-up has been computerized so that the number of flower visits, flight-activity and amount of nectar consumed by the bees could be automatically recorded. The results showed that the bees spend, on average, 244 minutes flying approximately 17 km each day! They consumed 180 microlitres of a 50% sugar solution every day, which was approximately equal to their own body weight (a microlitre is one millionth of a litre). They excreted about 136 milligrams of water each day, which is equal to the total body water of a bumblebee. In equivalent terms, a human weighing 70 kilograms would have to excrete about 50 litres of water per day. Professor Bertsch has concluded that water-loading, and not an energy scarcity, is the more important limiting factor in a bumblebee's life cycle. Before leaving bumblebees, it is interesting to note that they are frequently the first insects to emerge in spring when they fly at air temperatures of 2-3°C while maintaining body temperatures of 38°C, whereas Arctic mosquitoes stop flying below 8°C. Bumblebees, however, have an elegant biochemical mechanism to produce body heat, in addition to shivering. This consists of recycling a certain sugar continuously by breaking it down rapidly and then immediately re-synthesizing it. The process wastes energy but produces significant amounts of heat.

There are no bumblebees in Africa south of the Sahara and it appears that non-social carpenter bees (*Xylocopa* species) have taken their place. Together with Sue Nicolson of the University of Cape Town, I had the pleasure of studying the energy requirements of these fascinating insects on the slopes of Table Mountain adjacent to the city of Cape Town.

Like the bumblebees, carpenter bees have to process large amounts of watery nectar. For example, one species, *Xylocopa capitata*, visits *Virgilia* flowers, each of which contains only 0.7 microlitres of nectar, with an energy content equal to 1.9 calories. We found that a female carpenter bee had to obtain the nectar from 194 *Virgilia* flowers per hour or 3.2 flowers per minute to obtain sufficient energy to remain in continuous flight. They also had to visit about 1 678 flowers to provide sufficient pollen and nectar to raise one adult bee from the initial egg stage. In addition, these bees bore comparatively long tunnels into dead hardwood to provide nests for their developing young. Imagine yourself creating a tunnel about eight times your length and using only your teeth for the job. These bees have only enough energy to raise 3-4 offspring per year, which is most unusual in the insect world. Their offspring are rewarded by enjoying excellent protection within the tunnels during their development to adulthood.

Arizona honeybees

We move now from the lush vegetation on the slopes of Table Mountain to desert conditions in Arizona to study honeybees. One hot afternoon in Arizona, with air temperatures nudging 40°C, Neil Hadley, of Arizona State University, and I watched how honeybees were drinking from a small pond to ferry the water back to their hive. On reaching the hive they regurgitate the water, which then evaporates and cools the hive. Once within the hive, the bees whir their wings in unison to create an air stream through the hive to accelerate the evaporative cooling effects of the water.

We knew that the bees were loading large amounts of water into their crops, judging by their difficulty in becoming airborne, and we also knew that flying with these heavy payloads would require a very high metabolic rate to furnish their flight muscles with sufficient energy. This high energy demand would, in turn, require that the bees ventilate their respiratory tubules rapidly to supply their flight muscles with enough oxygen. We wondered if this increased ventilation rate under hot, dry desert conditions removed so much water from the bees by evaporation, that they would arrive at the hive dehydrated and be forced to use the water in their crops, originally meant for cooling the hive.

To find an answer to this puzzle, Neil and I first measured how much water the bees were capable of taking on board. It turns out that an 80 milligram bee can drink 50 milligrams of water or 65% of its body weight, then take off ponderously with much buzzing and fly back to its hive. The equivalent in human terms would be for an 80 kilogram person to drink 50 litres of water in a few minutes and then still be able to run, let alone fly. When we flew the water-loaded bees in our flight chambers, we found that they had one of the highest metabolic rates, relative to body weight, ever measured in nature; about 50 times the energy used by a sprinting human athlete on a relative weight basis. Simultaneously, the bees lost large amounts of water through evaporation but, and here is the beauty of the phenomenon, the bees manufactured just as much water from oxidizing sugar to supply the needed energy for flying, as they lost from evaporation. You will remember that when sugar is fully oxidized it yields carbon dioxide and water. Consequently, the water ferried back to the hive in the crops of the bees represented a net gain. These results also imply that under non-desert conditions many insects will go into a positive water balance merely by engaging in flight. The oxidization of either fats or sugars, in order to fuel the high energy demands of flying, would simultaneously produce large amounts of water. It appears then that a thirsty insect needs only to spin its flight motors to quench its thirst!

Shivering insects

Because insect flight usually requires a great deal of power and the flight muscles can only produce sufficient power when they are operating at a reasonably high temperature, many insects will shiver their flight muscles to build up sufficient heat before take-off. This is obvious in certain nocturnal moths but there are some interesting exceptions. For example, Bernd Heinrich of Vermont University found that a species of geometrid moth could fly under freezing conditions with a body temperature below 0°C. The wings are very large in this species in relation to its body weight and, consequently, the wing loading is low enough to obviate the need for a large power output by the wing muscles. Wing loading is the total weight of the insect divided by the surface area of the wings.

The above is not true for several species of dung beetle which exhibit body temperatures around 40°C while flying towards the dung pats left by large African mammals. The beetles are intent on carving out slices of fresh dung which they roll into balls. The balls are then propelled backwards by the beetles; they use their back legs to roll the ball while walking backwards on their front legs. When a suitable locality is reached they excavate a large hole in which to bury the dung ball. The eggs are then laid on the ball before it is covered with soil. The emerging larvae feed upon the dung; but the adults are more "fastidious" and suck out only the fluid portion.

The males, while swarming over a fresh dung pat, have an additional objective in mind, namely to capture a female's fancy. As a result, jousting matches ensue to capture the most desirable female's attention. Bernd Heinrich and Bart Bartholomew have studied these jousting matches, as well as the temperature regulation of the male dung beetles while the beetles were engaged in these contests. Bart, who is not an unusually tall man, has since revealed with considerable glee that it is the smallest male dung beetles that have the highest body temperatures, expend the most energy and win the most females.

The fly

This chapter on flight would be incomplete without reference to the insect whose brilliant aerobatic repertoire has earned it the eponym of "fly." No doubt, like most, your response to flies is one of disgust but this is certainly not

Professor Dethier's response. He has devoted a lifetime to studying the nutritional physiology of flies. One of his several texts, *To Know a Fly,* is an excellent example of how first class science can be communicated in a lucid and entertaining manner. In this delightful book, Dethier explains how flies can perform their impressive array of aerobatics such as half rolls, rolls off the top, stall turns and delicate landings by using their "massive" flight muscles and the greatly modified hind wings, known as halteres. The halteres act as gyroscopes to stabilize the fly's manoeuvres during flight.

Flies also enjoy excellent panoramic vision which allows them to take rapid action to avoid threatening predators. While in straight and level flight the visual world appears to the fly to float from front to back in a uniform manner. If it should now make a turn to the right, the visual world will rotate from front to back in the left eye and from back to front in the right eye. This way of seeing is known as rotational flow and if the fly should now pass a nearby object, the image of the object moves faster than the image of the background. The latter phenomenon is known as motion parallax and is an effective way of employing three dimensional vision. Next time you try to catch a fly, use both hands and approach it from both sides – a uni-lateral approach will seldom fool them. It is only just prior to landing that flies use binocular vision to execute their precise six-point landings.

What is the cost of all these impressive aerobatics? Dethier tells us that the blow fly, *Phormia*, which weighs about 25 milligrams, beats its wings at 200 strokes per second on a pleasant summer day (25°C) and uses 35.5 milligrams of glycogen per gram of body weight per hour. When *Phormia* is fed 1.8 milligrams of a fairly strong sugar solution (1 Molar), it can fly non-stop for a maximum of three hours. At the end of this period it will have used up 95% of its available energy reserves. These amounts may seem small, but in relation to the size of these flies, the energetic cost of flying is high. Perhaps to compensate for this, flies are rather lazy creatures, and Dethier informs us that female houseflies spend only 12.7% of their time walking or flying. It is only when hunger or thirst reaches a critical threshold that movement will be stimulated. In this regard, we could note that the taste sensitivity of a hungry fly for sugar is some ten million times greater than that of humans.

Similar data have been collected for other small flying insects and next time you are bitten by a mosquito, you may reflect on the fact that the mosquito *Culex pipiens* can vibrate its wings at almost 1000 strokes per second and fly 4.3 kilometres on only one ten thousandth of a gram of glucose. This amount sounds very little, but the fuel efficiency of mosquitoes is much lower than that of a jumbo jet. Many small insects pay a high energy cost to enjoy the benefits of flight.

Probably the smallest flying insects on our planet are parasitic wasps belonging to the families Trichogrammatidae and Mymaridae, or the fairy flies. They are only a fraction of a millimetre in length, yet in this tiny bundle, as small as the proverbial pin head, there is a fully developed digestive tract, respiratory system, hormonal and reproductive systems, as well as a nervous system which coordinates a perfect set of flight muscles. But perhaps more impressive is their complex behaviour; they are capable of seeking out the eggs of their hosts, which they parasitize by piercing the eggs with their ovipositors and then laying one or more tiny eggs. In addition, they are able to select eggs that have not been previously parasitized. Some species are even capable of swimming under water in search of their host's eggs, and it is difficult to explain how such small creatures can pierce the high surface tension of the water to gain access to these eggs. Nothing is known about the cost of flying in these miniature wasps, but their small size suggests that the cost would be very high.

By examining the cost of flying we have seen how important the ability to fly has been in opening up countless evolutionary opportunities for insects and birds. Not only are these animals able to fill unusual niches within an ecosystem, but they are ideally suited for exploiting a mosaic of widely dispersed nutritional resources. Imagine a bumblebee or honeybee that cannot fly – an impossible situation for both the bees and the plants they pollinate. Also, the ability to fly has allowed flying animals to undertake spectacular migrations and exploit extreme environments such as polar and desert regions, because of their swift and efficient means of locomotion. Consider, for example, the ability of Arctic terns to migrate from the Arctic to the Antarctic in a single season, a distance of over 11 600 kilometres.

Birds incapable of flying would be little more than feathered reptiles that keep their body temperatures fairly constant around 40°C. Most of the modern flightless birds, such as the ostrich and the emu, are large to compensate for their inability to fly. The reason for this is that in large animals, the energetic cost of walking and running is lower than in small animals. Why? Because large animals take fewer strides per minute and because they store more elastic energy in the tendons of their limbs. The tendons act much like recoiling springs and the release of their stored elastic energy lowers the cost of locomotion significantly. Why, however, are there no flying birds much smaller than two grams and none much larger than 19 kilograms? At the same time we could also ask: why are there no flying insects heavier than the rhinoceros beetles of tropical America (about 30 grams)?

Let us begin with the birds and try to answer the question of why none is smaller than two grams. If we compare the respiratory and blood systems of birds and insects, we discover that the transport of oxygen to the flight muscles of birds must travel along a more roundabout route via air sacs, the lungs, heart and blood vessels, than is the case in insects with the direct link between their flight muscles and the atmosphere via their respiratory tubes, known as tracheoles. This then may be the reason: that the avian respiratory system cannot deliver oxygen rapidly enough to support the extremely high metabolic rates which would be required for flight in birds weighing less than two grams. This is because the energetic cost of flight, similar to running and walking, increases with decreasing body size.

On the other hand, why are there no flying birds much larger than about 19 kilograms? The answer here is a little easier because as the volume of a bird increases, it increases in relation to the length x width x height of the body or, expressed more succinctly, to the third power (l^3). But the surface area of its wings, which determines the maximum wing loading, only increases in relation to the length x width, or to the second power (l^2). Hence the volume, which contains the mass of flying muscles, will, with increasing size, reach a certain critical weight, beyond which the surface area of the wings will no longer be able to provide sufficient lift. This critical limit appears to be somewhat more than the size of the largest flying birds, the trumpeter swan and the kori bustard, that can reach weights of 17 and 19 kilograms, respectively. In addition, very large birds have difficulty in manipulating their large wings close to the ground or water to give them enough lift during take-off. Consequently, they require long runways to become airborne and it is quite possible that the inability to reach take-off speed may be a more critical factor limiting body size in flying birds.

Perhaps a more interesting question to ask is: "Why are there no flying insects alive today that weigh more than 30 grams?" The simple answer is: we do not know. The efficient transport of oxygen to the flight muscles should not be a limiting factor, nor is the tough elastic exoskeleton, to which the muscles and wings are attached, unsuitable for flight. It is, in fact, in many ways an ideal airframe. The answer to our question is probably not based on any physiological phenomenon. For example, although it appears to be physiologically possible for, say, a 5 kilogram rhinoceros beetle to fly comfortably, the very large larvae of such a beetle may not be able to avoid predation, particularly during moulting when they would be most vulnerable. Other complex ecological reasons could also be involved but it remains an excellent subject for speculation.

The last part of our discussion has obviously been an oversimplification; there remain several other important factors, such as the density and oxygen content of the air and the force of gravity, to be considered. There is some evidence to suggest that these factors may have varied significantly during the history of our planet, and it is possible that the limits of body size for flight have also varied concurrently. This may explain how giant dragonflies with wingspans as great as 70 cm were able to fly during the Upper Carboniferous and the Upper Permian, between 230 and 280 million years ago. Also, we are told by palaeontologists that giant flying reptiles, known as *Pteroanodons*, had wingspans of almost eight metres (25 feet). Eighty million years ago they used these magnificent skin wings to soar on swiftly rising thermals above what is now the state of Kansas.

Dare we hope for their return?

The
Cost
of
Swimming

The term "swimming" describes many different patterns of locomotion through water, from the delicate extension of the false legs (or pseudopodia) of single-celled amoebae to the elegant sprinting of streamlined dolphins. Between these extremes countless other patterns exist, such as rowing in water beetles, jet-propulsion in squid, the wriggling of marine worms, surfing in certain marine snails and the torpedo-like swiftness of swimming penguins. Let us, for the sake of brevity, limit our discussion to swimming in fish.

Fish have existed on our planet for over 450 million years. They have exploited the advantages of the vertebrate body plan by filling the many available aquatic niches in both fresh and sea water. Today there are many thousands of species of every conceivable shape, colour and size, ranging from 1.5 grams to 4 000 kilograms. One need only compare the life-style of the small pup fish that survive water temperatures of 43°C in Death Valley, California, with that of the ice fish of the Antarctic, that survive subzero temperatures but die when warmed to 8°C, to appreciate the wide adaptive radiation that fishes have undergone.

Humans have exploited fish as an important food resource since time immemorial. The reason they have been an integral part of our diet for so long, apart from their obvious nutritional value, is because they are reasonably easy to catch and the energy they consume is free. We do not have to plough and fertilize fields to feed them – the ocean provides everything. In addition to this great advantage, fish are efficient users of the energy they consume, and many of them feed upon plants and organisms which are useless to us. For example, mullet suck up the mud from the bottom of estuaries to obtain their food, yet when grilled over an open fire they make an appetizing dish. Also, many fish digest their food efficiently; the efficiency will vary with the nature of the diet but can approach as much as 90 percent. Swimming is the most efficient means of locomotion, which means that fish waste comparatively little energy in foraging for their food, thereby increasing their overall efficiency of energy use.

Sadly, however, we have over-exploited the fisheries of the world to such an extent that it now costs five times more energy to catch and deliver a tonne of fish from the ocean than the amount of energy this weight of fish will provide in our diet. Many fisheries have already collapsed and the failure of the Newfoundland fishery off the east coast of Canada is one of the major ecological disasters of this century.

We begin our study of the cost of swimming by asking why swimming is such an efficient form of locomotion. There are several reasons for this, but first let's find out how fish swim. While swimming, a fish uses the powerful muscles in the tail to beat its tail from side to side. This action forces the water away to either side of the fish and the resultant force propels the fish forwards. Possession of a large tail should therefore be an advantage and experiments, in which tail size was artificially altered, have confirmed this. The major reason for this large difference is that fish, unlike terrestrial animals, do not have to support their own body weight; the high density of the water provides them with buoyancy which largely cancels out the effects of gravity. The latter is of course a major burden on running mammals and birds engaged in flapping flight.

The high density of water, however, has the disadvantage of producing friction or drag forces, which slow fish down when they swim. To overcome this, many fish have developed highly streamlined shapes, similar to marine mammals such as dolphins and seals (a good example of convergent evolution). Also, the relative importance of drag forces is reduced in large fish. This is because the larger a fish becomes, the smaller its relative surface area becomes (remember the example of the football and the marble). This means that drag per unit body weight will decrease with increasing body size, because drag is proportional to surface area if speed remains constant. In addition, the larger a fish becomes, the greater the mass of the swimming muscles becomes, and consequently their power output becomes greater as well. In this way larger fish enjoy a greater power output to drag ratio than small fish. It is not surprising, then, that the fastest fish in the ocean, such as marlin and the great white sharks, are both large and elegantly streamlined.

The rate at which chemical reactions provide energy in animal tissues depends largely on the temperature of the relevant tissue. We have already seen that many flying insects increase their body temperature by shivering to meet the high energetic demands of flying. But what do fish do in this regard?

Most fish adapt to the temperature of the surrounding water. This is what we would expect as they are all so-called "cold-blooded" animals. When the water flows over their gills, the relatively small amount of heat generated from muscular contraction during swimming rapidly dissipates from the blood through the thin gill membranes to the surrounding water. This loss of heat would be a distinct embarrassment to deep-water hunters, such as the swift tuna, when they encounter a cold current. They would slow down to such an extent that their hunting efficiency would be seriously impaired.

To prevent this happening, these swift swimmers have evolved heat exchangers between their muscles and their gills. These heat exchangers consist of networks of fine blood vessels in which the small veins, carrying the warm blood from the muscles, are arranged in such a way that this warm blood is brought into close contact with the small arterioles carrying cold blood from the gills to the muscles. Because the flow of the venous and arterial blood is in opposite directions, heat is rapidly and efficiently transferred from the warm blood draining from the muscles to the cold arterial blood supplying the muscles. This constitutes a thermal shunt, as a large portion of the heat developed in the muscles is shunted back to the muscles by transferring the heat from the venous to the arterial vessels. Engineers refer to this type of shunt as a counter-current heat exchanger.

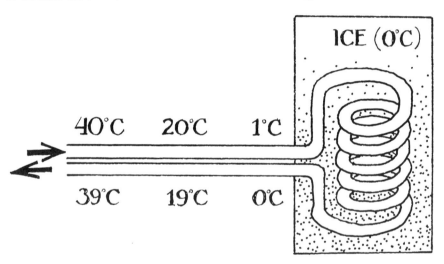

Model of a counter-current heat exchanger such as might exist in the legs of an animal standing on ice. In this case, heat is conducted from the incoming water to the outflowing water, so that in the steady-state condition the outflowing water is pre-warmed to within 1°C of the incoming water. From Schmidt-Nielson (1972).

Heat conserved in this way allows the tuna to maintain a body temperature some 14°C above that of the surrounding water and to enjoy the benefits of a high cruising speed, even when unexpected changes in water temperature occur.

We should therefore be cautious about describing tuna as cold-blooded, and this is equally true for larger and even more impressive hunters, such as marlin, that have evolved special adaptations to keep their brains and optic nerves warm. Situated close to the brain of these splendid hunters is a mass of modified eye-muscle tissue, packed with tiny sub-cellular structures that are responsible for much of the energy turn-over in these cells. These small structures are known as mitochondria. Normally, when mitochondria break down nutrients to provide energy for muscle contraction, they produce some heat as well as a considerable amount of an energy-rich compound known as adenosine triphosphate (ATP). The ATP then transfers energy to the contracting muscle filaments.

In the specialized eye-muscle tissue of the marlin, how-ever, the turnover of ATP is accelerated and consequently much more heat is produced. This heat keeps the tempera-ture of the optical apparatus well above that of the rest of the body, thereby maintaining a high speed of conduction of the nerve impulses travelling along the optic nerves. This en-sures that these magnificent predators retain maximum visual acuity at all times to locate and tackle their prey – their only source of energy.

In contrast, several species of polar fish have body temperatures well below the freezing point of their tissue fluids. They exist in what is known as a "super-cooled" state and are quite safe unless they make contact with ice, whereupon they immediately become frozen and perish. This, naturally, seldom happens. The blood of Antarctic ice fish contains no haemoglobin because of their inactive life-style and the high solubility of oxygen in cold water.

Catching fish with a throw net in tropical waters is one of my more pleasant memories. It takes a great deal of practice to learn this skill, but the excitement of drawing in the net makes it all worth while. As the net is drawn ashore, the silver flanks of the fish flash in the sunlight and hereby hangs a tale. The shiny silver appearance of so many kinds of fish is due to the presence of an extremely large number of guanine crystals, about a million per square centimetre, which are deposited in the fish scales. The crystals of this nitrogenous compound are laid down in a specific pattern so that light falling on the side of the fish is reflected, but light rays from above will largely pass between the crystals. This makes the fish almost invisible to predators from above but, when seen from the side, they are clearly visible.

Stephen Downes, in his beautifully illustrated book *The New Compleat Angler*, tells how a solution of fish scales was used in the last century to coat small glass spheres in order to produce the first artificial pearls. When one remembers, however, that guanine is one of the major components of marine bird droppings, it detracts somewhat from the romantic image of the pearls.

The flashing silver sides of fish are important for communicating signals, particularly when they are swimming in schools, yet there does not seem to be consistent leadership in fish schools. Next time you watch a school of fish, you will notice when they swerve to, say, the left, the former leaders are now on the flank and those that were on the flank now assume the lead. Schooling also provides some interesting biophysical questions with regard to the distance between fish within a school. If fish swim too closely to one another, they will swim within a zone of turbulence created by their nearest neighbours, and this means that the smooth laminar flow of water over their own bodies will be disturbed and their efficiency of locomotion will be compromised. On the other hand, fish can actually coast for short distances along the edge of the turbulent vortices created by their neighbours, thereby using energy expended by the fish swimming in front of them.

Schooling has other advantages, not least of which is improved predator evasion. When a large predator strikes, the ranks of the fish may close and confront the predator or, more commonly, the school will explode in all directions, confusing the predator to such an extent that it abandons the chase. Schooling can improve foraging efficiency, particularly by improving the probability of finding a source of food, but it may also increase competition once the resource has been located. Schooling will under certain conditions also facilitate the finding of mates and improve the efficiency of fertilization in spawning species.

Although schooling occurs in many fish species, there are no truly social species of fish as there are among the termites, wasps, bees and ants. The latter species are, in present day jargon, described as *eusocial* because they exhibit cooperative care of the young; there is an overlap between generations; and fertility is suppressed in all females except in the queen caste. Eusociality in animals can, as we shall see later, reduce the energetic cost of living and the absence of eusocial animals in aquatic systems presents a fascinating evolutionary puzzle. Possible reasons for this are that it is more difficult to navigate back to the communal nest when underwater and that it is difficult to build nests in water and ventilate them with sufficient oxygen. Alternatively, there may be no need for eusocial animals in aquatic systems, because conditions in the ocean and large lakes are either sufficiently stable or predictable so as not to require the flexible lifestyle provided by eusociality.

The closest example to eusociality is found in clown fish that live among the tentacles of anemones. These small communities of brightly coloured fish are immune to the anemone's toxins and scavenge for food particles in and around the anemone. Within this small group of fish there is always a dominant female that is fertile and sexually active. The remaining females are infertile, but should the dominant female die, another dominant female will emerge from the small family and become sexually active. Clown fish are therefore social, but cannot strictly be described as eusocial.

Although swimming is an energy-efficient form of locomotion, we must remember that the transport of oxygen to the muscles of fish is a rather inefficient process when compared to mammals and birds. This is because oxygenated blood from the gills of fish, unlike in other vertebrates, does not return directly to the heart from where it can be pumped under pressure rapidly to the muscles. Instead, it drains slowly under low pressure from the gills directly to the muscles. For this reason, trout can swim indefinitely at speeds of about 0.6 metres per second; but short bursts of speed at 2.8 metres per second will rapidly exhaust its oxygen reserves.

A good example of this is provided by a fly fisherman angling for trout. When the trout takes the fly it may momentarily explode above the water, before swimming strongly away. Adrenaline is now peaking in both the trout and the fisherman. The fisherman, however, is a "good sport" and is using a very fragile line which requires him to play the fish slowly until it lands exhausted, belly up, at his feet. I have analyzed the blood and tissues of trout in this exhausted state, and found that fish in this condition had accumulated a large oxygen debt with very high concentrations of lactic acid in the muscles. Potassium levels in the blood were also high and all these signs are typical of stress and exhaustion, but not necessarily of experiencing pain. The latter remains a contentious issue. It is possible, nevertheless, that when we eat a fish that has been played to exhaustion, the taste may be adversely affected by the above changes in muscle chemistry. This has not been conclusively proven. In any event, gentle poaching in a good white wine with suitable herbs should compensate for any aberrant biochemistry!

Fish, therefore, are forced to shift to anaerobic (without oxygen) metabolism when faced with prolonged, sustained activity beyond their natural endurance. These endurance limits are far lower than those of most mammals and birds. When this happens lactic acid accumulates in the tissues until the oxygen debt is paid back. However, not all fish use this metabolic pathway when they over-exert themselves or are forced to survive in oxygen depleted waters. For instance, certain cyprinid fish, such as carp, have the ability to produce energy-rich ATP by metabolizing sugar into ethyl alcohol when faced with oxygen stress. The alcohol level in their bloodstream never rises above our regulation driving limits, because alcohol diffuses freely across the gill membranes into the surrounding water. In this way the rather mournful looking carp compensates for the lack of oxygen but remains completely sober.

Apart from the rather inefficient transport of oxygen from the gills to the muscles of fish, these animals are faced with an additional problem which prevents them from achieving really high levels of sustained energy output: the high density of the medium (water) which they "breathe." The density of the water irrigating their gills is 1 000 times that of air and it must also be remembered that water usually contains less than one percent dissolved oxygen. In contrast, the high concentration of oxygen in air (21 percent) and its much lower density has made it far easier for air-breathing vertebrates to ventilate their respiratory surfaces. No surprise, then, that warm-blooded vertebrates with an energy turnover some 13 to 17 times greater than cold-blooded vertebrates are all air breathers. There is, however, a considerable number of air-breathing fish, such as the famous African lung fish, which does not have a high energy turnover but is capable of breathing air. It seems hardly to have changed at all during the last 400 million years. Hence the nickname of a living fossil.

We have not even scratched the surface of fish biology in this brief essay. A rich and rewarding literature awaits the serious reader of this subject and the suggested reading at the end of this volume would make a good start. Better still, become a fisherman, then you may also become a philosopher.

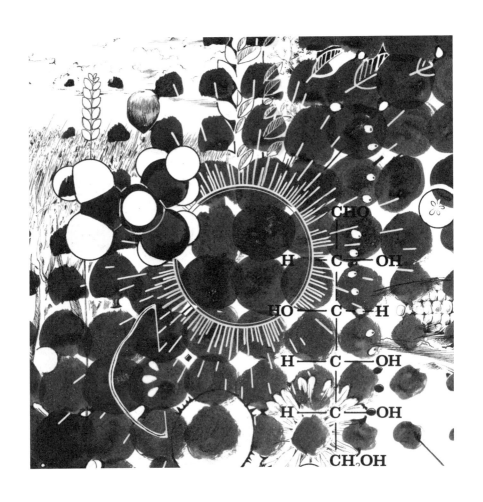

Fuels
for
Beating
Entropy

Animals have evolved an astounding variety of ways in which to obtain energy from the environment. Mites, smaller than a pinhead, scavenge through your bed clothes feeding on fragments of shed skin; others live in the hair follicles of your eyelashes feeding on fatty secretions. Sharks that have been disemboweled swim back and feed upon their own viscera. Minute book lice live for generations in the dry dust of old libraries, feeding on the old-fashioned glues used to bind books. The maggots of blowflies feed on rotting flesh and were used until fairly recently to clean up septic wounds in human patients. Some female spiders consume their male partners during mating, while others allow their newly-hatched offspring to eat them to prevent cannibalism among their young. Many marine snails filter out microscopic food particles from the sea water they pump through their digestive systems. Similarly, huge baleen whales survive by filtering out small crustaceans from sea water washing through their cavernous maws.

Mosquitoes obtain sufficient energy from a blood meal, measured in millionths of a litre, whereas an elephant requires tonnes of crude plant material for its massive bulk. Nevertheless, once the digested nutrients have been absorbed and transported to the cells, the functions they serve and the changes they undergo are very similar in both the mosquito and the elephant and, for that matter, in most other animals as well.

The greatest variation in animal nutrition lies in the feeding processes. These include sucking, absorbing, grinding, cracking, husking, biting, tearing, filtering and even using poisonous venoms or violence to capture food items. This variation is really a reflection of how specialized animal species have become in their feeding habits and how remarkably little overlap exists among the many nutritional niches they occupy. To put it more bluntly, "feeding a tiger apples will not turn it into a vegetarian," and we should refrain from feeding wild animals junk food in zoos and parks.

Some higher primates, however, particularly baboons and humans, use a wide variety of foods and invade the nutritional niches of other animals; but cultural preferences among humans still play an important role in this respect. Cannibalism, for example, is a strong taboo in most societies, as is the eating of dogs and horses in the English-speaking world. Some Japanese take great risks when eating puffer fish (fugu), from which most of the highly toxic tissues have been removed by specially licensed chefs. Deaths from the puffer toxin are still common and one can only wonder about the gourmet rewards (a slight tingling sensation of the lips) that justify such risks. Vegetarianism is an integral part of several religions; it is also popular among some dedicated environmentalists, who realize that it takes about six to ten times as much energy to produce a kilogram of meat than is required for the same amount of vegetables.

We require energy every second of our lives to maintain vital life processes. Even when we are at complete rest in a comfortably warm environment, we require energy for respiration, for our hearts to beat and to maintain a measure of muscle tone. If the temperature should fall below about 20°C and we do not have access to warm clothing, we shall start to shiver and our energy requirements will increase markedly. The same is naturally true for reproduction and lactation. A human pregnancy will, in total, require an additional energy allowance of 40 000 kilocalories, and a mother producing 850 millilitres of milk per day will require about 750 kilocalories extra per day.

Activity is one of the major factors influencing energy requirements; a small bird requires about 17 kilocalories per day while resting, about 30 kilocalories per day when engaged in courtship and territorial defense, and as much as 103 kilocalories per day during distant flight. Similarly, a large ruminant like a moose will increase its resting metabolic rate some 1.59 times when foraging for food, about 3.0 times when playing and will show an increase of more than 8.0 times the resting rate when running.

Much of the energy we consume is, however, used to repair our tissues which are continually being broken down. Take a good look at your neighbour – in about 80 days time almost 90 per cent of his body chemicals will have been replaced with new molecules. Fortunately you will still recognize him, as his DNA ensures that he will be re-synthesized into his familiar shape and size. Yet, in spite of all life's cunning tricks, which the famous French entomologist, Henri Fabre, described as *le savant brigandage de la vie*, life must ultimately obey the laws of physics. They naturally include the first and the second laws of thermodynamics which state (1) that energy is neither created nor destroyed and (2) that the disorder or *entropy* of a system and its surroundings always increases. This means that the exquisite organization of the molecules, cells and tissues in animal bodies is threatening to decay continuously into a state of disorder or increasing entropy. This is why some physicists have described *life* as a process of beating entropy, at least temporarily.

Almost all the energy we consume is ultimately derived from the efficient process of photosynthesis in plants. Although plants trap only about 15 per cent of the incoming sunlight, the energy they do capture is processed very efficiently. In fact, one photon of the sun's electromagnetic energy is captured with 100 percent efficiency to release one "high-energy" electron which, together with countless additional electrons released in a similar manner, trigger a chain of biochemical reactions in the plant that eventually produce complex molecules. When we eat plants or the herbivores that live on plants, we are harvesting the energy transferred by these electrons. This is used to fuel the many synthetic reactions required to repair our tissues and fight entropy.

An alternative (non-solar) source of energy, available to certain animals, is provided by symbiotic bacteria which oxidize sulphides to elemental sulphur. This, for example, occurs in huge pogonophron worms that live at great depths on the sea floor close to hydrothermal vents.

How efficiently do animals use the energy they consume? There are many ways of measuring this efficiency involving elaborate gas analyses, nutrient balances and radio-isotope studies. Perhaps the simplest and most informative way is to express efficiency as the percentage of the consumed energy (C) which is actually deposited in the tissues (D). Therefore efficiency (as %) = D/C X 100. Studies of this kind show that terrestrial detritus eaters, like earthworms, which turn coarse compost into fine compost, have a very low efficiency of about six percent. Terrestrial herbivores, for example antelope, range from 13-20 percent and terrestrial grain eaters, such as certain birds and rodents, have values of about 24 percent. Among the highest D/C values recorded to date are for terrestrial carnivores, like lions (46 percent), and young mammals still dependent on milk (43 percent). These differences are largely a reflection of the quality of the food consumed, particularly the amount of digestible energy available in the food.

The lifestyle of the animal also plays a role in determining D/C values; those involved in continual high activity will have a lower value. This is why warm-blooded animals, such as birds and mammals that engage in a great deal of sustained activity at high body temperatures and often at high speed, have a cost of living which can be as much as thirteen times higher than say a reptile, like a crocodile, which can exist for many weeks without feeding. One of the very highest

D/C values reported is 54 percent for carpenter bee larvae. This is understandable when one realizes that the larvae of these bees are confined throughout their development to a small cell within a dead tree trunk. They feed on a high-energy mixture of nectar and pollen – food for the gods.

Let us now take a closer look at the fuels which supply energy to animals. These can be categorized as proteins, carbohydrates and fats.

Proteins

Most people would claim that proteins are essential in the diets of all animals and definitely so in the case of humans. They would not, however, be completely correct because proteins, as such, are not required, but rather certain amino acids which are contained in proteins. Amino acids are the building blocks of proteins and the amount and kind of the amino acids contained in a protein determine its quality.

Every animal species tested to date has a specific requirement for amino acids. These are known as essential amino acids because the animal in question cannot synthesize them. The extent to which a protein supplies these essential amino acids is then a measure of its quality. For example, in human nutrition the protein keratin, which is the main constituent of horns, nails and hair, lacks certain essential amino acids and we would not survive on keratin as a sole source of protein. The proteins with the highest quality are those contained in egg yolks, in milk proteins (casein) and the proteins in the tissues of various animals, like fish, lobsters, poultry and the large mammals.

A perfectly balanced vegetable protein does not exist and for this reason strict vegetarians (vegans) must have fairly expert knowledge of how to mix and match the various constituents of their diets, so that the deficiency in one food is counterbalanced by an adequate supply from another source. For example maize (corn), which is a basic high-energy food for millions of poor people in Africa and parts of South

America, is deficient in the amino acid lysine and the B-vitamin niacin. Humans cannot survive on maize alone. Young growing children rapidly develop severe malnutrition symptoms, including pellagra, on such a diet. But when maize is mixed with beans, the amino acids from both sources are integrated to provide an adequate spectrum of essential amino acids.

This mixing and matching of high yielding, high-energy crops, usually deficient in one or more amino acids, with either good quality vegetable or animal protein, has taken place all over the world for thousands of years. Each culture has through time developed the most practical mixture. In this way East Indians mix rice with lentils; the Japanese mix soya beans with rice or, when available, fish with rice.

The importation of potatoes from South America to Europe in 1570 led to this crop rivalling wheat as the main source of nutritional energy in Europe. Both potatoes and wheat lack high quality proteins and are usually balanced by animal products such as meat, cheese or fish in Europe. In the past, when animal products were not available, peas and lentils were used. During the long northern European winters, the requirement for B-vitamins was mostly satisfied by drinking beer, while vitamin C was largely supplied by sauerkraut. Hence the humorous designation of this extensive area as the "beer and sauerkraut belt" – hardly flattering to one of the so-called centres of Western culture!

Although all animals require specific amounts and kinds of essential amino acids for their cells to function normally, in certain cases these are not actually required in the diet. In certain insect species such as termites, the microbes inhabiting the digestive tract synthesize amino acids for their own use and these are eventually passed on to the host animal. Similarly, in the large forestomach of animals that chew the cud, the millions of teeming microbes are able to synthesize amino acids from simple nitrogen compounds such as urea. This means that in times of need, for example during a long drought on the African savannah, when the protein content of the natural pasture has fallen to critical levels, ruminant animals can divert the urea, normally excreted by the kidneys, to the forestomach. Here it is recycled into microbial protein of a high quality. In this way a desert camel can produce nutritious milk for its offspring from the sparse, and crude dry plant material of the Sahara.

The functions of amino acids in animal tissue are very diverse. They are used to synthesize structural proteins such as muscle, hair, feathers, horns and hooves as well as for the synthesis of complex enzymes and hormones that integrate and control many vitally important biochemical reactions within animal cells. For this reason, a deficiency of any one of the essential amino acids has serious consequences and will eventually lead to death. Also, amino acid deficiencies are the most likely deficiencies to occur, particularly in steppe and savannah areas that experience long periods of drought.

Deficiencies can also occur in advanced countries among educated people. This is partly due to the poor ability of humans to develop taste preferences for nutrients that are deficient in their diet. Our major craving is for calories (sweetness), a little salt and in some cases animal fat. Consequently, many middle-aged single people living alone in London were found to exist mainly on buttered toast and sweet tea. It satisfied their energy requirements and was easy to prepare. They eventually suffered from severe malnutrition, which became known as the "tea and toast syndrome."

An important function of proteins, or more correctly amino acids, is to provide the raw materials for the yolking of eggs in egg-laying animals. In the case of mosquitoes, the female must first obtain sufficient protein from a blood meal before she can start to yolk her eggs prior to egg-laying. Different species of mosquito have evolved specific amino acid requirements for this purpose. These are related to the amino acid make-up of the blood of their natural hosts. For example, human blood has a low concentration of an amino acid known as isoleucine, and this may be the reason why female mosquitoes of a certain species are forced to seek multiple blood meals from humans for a single egg-laying cycle. This has serious implications for the spread of disease, particularly under crowded sleeping conditions.

Proteins are widely distributed in nature. Fresh young plant leaves contain adequate amounts of proteins for most herbivores. As we have seen, cereal grains are deficient in certain amino acids when used for human nutrition but, when mixed with the seeds of legumes such as peas and beans, which are high in protein, an adequate spectrum of amino acids is obtained. Even the vegetative portion of many legumes is comparatively high in protein, and the best example of this is probably alfalfa or lucerne hay. It is relished by most herbivores, but the reason for its high palatability is not yet known. Most animal products provide an excellent quality of protein and, for this reason, small quantities mixed with cereal grains can make a significant difference to the nutritional quality of a diet. Anyone who has been to an English boarding school or in the army will remember the old standby of macaroni and cheese.

Cheese has today become somewhat of a luxury by worldwide standards, but it must have been known to the earliest pastoral nomads that milked their ruminant animals thousands of years ago. They would have noticed very early in the development of animal husbandry that milk coagulates in the "milk" stomach of kids and lambs to form curds and whey. From this observation it would have been a comparatively easy step to use the rennin from these stomachs to make cheese from their surplus milk. The volume of cheese is much smaller than the original milk volume from which it has been made. It is also drier and more acidic and for these reasons it is far more suitable for storage and transport.

An alternative method of preservation of milk was practiced by the famous Tartar horsemen who, after milking their mares, dried the milk for later use on their sweeping raids into central Europe. In contrast, modern cheese making is a complex process with many secrets and closely guarded bacterial or fungal cultures used to ripen and flavour the many varieties of cheese. The pinnacle of achievement in cheesemaking is probably found in English Stilton, with its firm but creamy consistency and the unusual flavour imparted by the blue-green penicillium mould; although many Italians would insist that this prize should be awarded to aged Parmesan cheese.

Finally, we should not lose sight of the important function of proteins in providing energy for a variety of living processes, including the beating of entropy. Once a protein has been reduced to its constituent amino acids, all that is required is the removal of nitrogen from the amino acid to produce an organic acid. The latter can then be oxidized in the normal way to provide energy, carbon dioxide and water. This is what happens in the case of carnivores that exist on an almost pure protein diet, with very little access to carbohydrates. It would also apply partially to the Inuit living in the far north of Canada, who existed largely on meat products during the winter, before the invasion of Western dietary customs. In view of the high fat content of Arctic mammals, their fat intake would, nevertheless, also have been high.

Carbohydrates

Walk through a lush tropical rain forest, across the prairies or the African savannah and you are surrounded by vast quantities of carbohydrates, because they are the most common plant products on our planet. The simplest of these will be sugars such as glucose, fructose and sucrose. The major complex carbohydrates will be starch and cellulose. These are made up of long chains of glucose molecules, but there is a very important difference between starch and cellulose in the way the glucose molecules are joined together. In the case of starch, they are joined by the so-called α-linkage which is easily digestible, whereas in cellulose the linkage is of the β-kind, which very few animals can split. Those animals that have cellulose-rich diets usually possess an enlarged portion of their digestive tracts, in which the passage of the food mass is slowed down and where the cellulose is attacked by microbes. These microbes possess the enzymes to split the

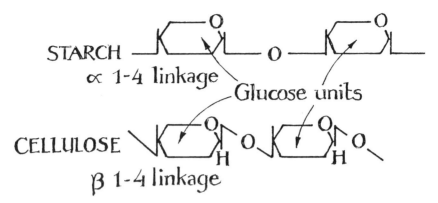

STARCH

α 1-4 linkage

Glucose units

CELLULOSE

β 1-4 linkage

β-linkage and release the glucose molecules. Humans are poor digesters of cellulose and the high-fibre breakfast cereal you may be eating for health reasons is not too different from the cardboard container in which it is sold.

The sugars that are released from the digestion of complex carbohydrates are instant fuels for many activities in the animal body. They provide energy for maintaining the correct electrical charge across cell membranes; for synthesizing complex biochemicals; for the conduction of nerve impulses and for supplying fuel for muscle contraction. It is also important to note that the brain can only use glucose as a fuel. Psychologists at University College Swansea have shown that glucose consumption speeds up brain activity, enhances memory and speeds up reaction time.

Excess sugars in the body are stored as glycogen, a form of animal starch, or as fat. The amount of stored glycogen in most animals is, however, limited and marathon runners use up this store fairly early in the race and are then obliged to switch to body fat to fuel muscle contraction. Many athletes, therefore, stoke up on pasta and other starch-rich foods on the day before the race. They would, perhaps, be just as well advised to increase their fat consumption at the same time, because body fat is more easily synthesized from dietary fat than from carbohydrate, as any overweight gourmand will testify.

Cellulose, as we have seen, is very widely distributed because it is the main structural component of plants. In contrast, high-yielding sources of the digestible carbohydrates, like starch, are restricted to cereal grains, sugar cane, sugar beets and root crops such as potatoes, sweet potatoes and cassava or manioc. The highest yield of digestible calories per acre are obtained from sugar cane and maize (corn). Although under certain conditions, algae growing in shallow ponds can out-perform the cereal crops in this respect.

Wheat is still the main source of digestible carbohydrate in North America and Europe. Not only is it a rich source of digestible energy but its protein content and quality is fair. Also, wheat protein (gluten) has the property of becoming both plastic and elastic when ground up and mixed with water. This property results in a dough which traps the carbon dioxide gas produced by the yeast; the loaf can then expand in volume without rupturing, making the universally popular "leavened" bread possible. Rye is easier to grow than wheat and, although its protein has similar properties to the gluten protein of wheat, it is not nearly as efficient in producing high quality "raised" bread. It was perhaps for this reason that rye breads became a staple food for the working classes in the Middle Ages and wheat consumption was restricted to the wealthier classes.

Pure sugar is such an everyday commodity today, it is difficult to imagine a world without it. Yet, until the 18th century when sugar beets were first cultivated extensively, most people had to rely on certain fruits and honey for a source of sweetness. The production of raw sugar from sugar cane in India dates back to 500 B.C., but this source was restricted to the privileged few. Sweetness is a super taste stimulus for humans. Those dieticians who would like to remove it from our diet should ponder on our long association with sugar, even prior to the Stone Age when we were gatherers of sweet fruits. The plants that produce sweet fruits benefit from the dispersion of their seeds and it is possible that our predilection for sweetness may date back even further to our tree-dwelling primate ancestors.

In spite of the intense sweetness of certain fruits, particularly dates which some believe to be the oldest cultivated crop, they are no match for honey. Honey has probably been a prized luxury since the earliest humans emerged on the African savannah, and the present-day close relationship between honey guiding birds and humans in Africa must have taken a long time to evolve. Recent research in Kenya by Professors Isack and Reyer have now confirmed that the Boran people of Kenya can attract the attention of these birds by special whistling noises. They can then deduce how far and in which direction the bee colony is, by observing certain peculiarities in the bird's flight pattern. When the honey gatherers arrive close to the colony, the bird's behaviour changes again and they begin then to search carefully for the nest.

Once the nest is found, the gatherers smoke out the bees to rob the hive of honey. The honeyguide patiently waits its turn to devour left-over honey and beeswax. It is one of the very few animals capable of digesting wax and the only bird species able to do so. This was first discovered by missionaries in East Africa, who noticed how the honeyguides fed on the beeswax candles when services were conducted in open-air churches.

Isack and Reyer also believe, from the results of their controlled field experiments, that the honeyguides have knowledge of the position of a particular bee colony before commencing a guiding operation – a remarkable example of rational behaviour in an animal.

Honey is referred to in the very earliest records of Egyptian hieroglyphics, as well as in the Old Testament, and in the classical literature of Greece and Rome. It is surprising, therefore, how long it took for us to understand the relationships between nectar production by flowers and the production of honey by bees. In classical times it was believed that bees manufactured honey from a special dew which fell from the skies, described by the famous Roman naturalist, Pliny, as "a sort of saliva of the stars." Today, of course, we know that honey is made from nectar which consists mostly of the three sugars: sucrose, fructose and glucose. The sugar concentration of nectar depends on the flower species from which it came, the time of day and the climatic conditions. Recorded values range from seven to more than fifty percent sugar. The source of the nectar also influences the flavour of honey, which can range in colour from a pale golden to a tarry black at certain times of the year in the African savannah.

Harold McGee, in his book *On Food and Cooking*, has calculated that for every 500 grams of honey taken to market from a hive containing 20 000 female workers, 4 000 grams are used by the hive for its own maintenance. He then estimates that the total distance which must be flown by a bee to gather enough nectar for 500 grams of surplus honey is equal to flying three times around the earth. A bee's fuel consumption, according to McGee, is about seven million miles to a gallon of honey.

Before leaving the carbohydrates, we should look at one of their major fermentation products, namely ethyl alcohol. Alcohol, when metabolized, yields energy which can even contribute to one becoming overweight, and it must therefore be classified as a food. Its metabolism is not, as most people think, speeded up by vigorous exercise. This is because the rate of alcohol "break-down" in our tissues is governed by the amount of an enzyme known as alcohol dehydrogenese. The availability and concentration of this enzyme is not influenced by exercise and it is, in any event, usually in short supply. Hence the frequency of recriminations after the "night before."

The first discovery and use of alcohol by humans is lost in time but one can imagine how easy it would have been to make such a discovery. During a warm summer, a surplus of sweet fruit is left in a container and overlooked for a week or two. The owner returns, drinks the fermented juice from the fruit and within 30 minutes or so, he feels braver, the world is after all not such a bad place. If these pleasant feelings are not compromised the following day by a hangover, *Homo sapiens sapiens* would surely have put two and two together and repeated this experiment. Today the manufacturing of both wine and beer have become high-technology industries, employing advanced chemical and genetic engineering techniques. Nevertheless, they still depend on the same basic reaction of yeasts turning sugars into alcohol by simple fermentation to produce good beers and fine wines.

Fats

Generally speaking, plants produce oils and most vertebrate animals synthesize fat, with the exception of many oily fish species. Both fats and oils consist essentially of fatty acids bound to simple alcohols. Fats, however, have higher melting points than oils because their constituent fatty acids are longer and more fully saturated with hydrogen atoms.

Margarine is made from vegetable oils. The process consists largely of adding hydrogen atoms to the fatty acids in the oils, using a suitable catalyst. This raises the melting point of the oils, which now take on a similar chemistry and consistency to butter. There are, therefore, very few, if any, advantages for our health when we replace butter with margarine.

Fat, as such, is not essential in our diet but it is a highly prized flavouring agent in the cuisines of most cultures. In addition, we require three essential fatty acids in our diet, namely, arachidonic, linoleic and linolenic acids. Animal fats are rich in these fatty acids and therefore a convenient source. Vegans should be alert to the danger of developing deficiencies of these important acids. Conversely, diets high in animal fats have been implicated in coronary heart disease, and standard medical advice today is to make only moderate use of animal fats.

The major function of fats is to provide a light, concentrated compound for storing energy in the animal body. Fat contains about 2.25 times more potential energy than the other tissues of the body and requires correspondingly more energy for its synthesis. When required to provide energy, the fatty acids are chopped up into hydrocarbon fragments and the hydrogen atoms are stripped sequentially from these fragments, thereby transferring electrons to energy-rich compounds, which then provide energy for countless biochemical reactions.

When fats are fully oxidized, they naturally produce carbon dioxide and water and, because of the large number of hydrogen atoms in fat, they produce considerably more of this metabolic water than proteins and carbohydrates do when they are fully oxidized. Fatty tissue can therefore act as an important form of water storage and many desert animals are able to survive exclusively on this source of water.

A specialized fatty tissue, known as brown fat because of the colour imparted by the presence of special pigments and its rich blood supply, has a metabolic rate some 20 times that of normal fatty issue. This extremely high metabolic rate results in the production of large amounts of heat and is commonly used by hibernators to raise their body temperatures in spring, without resorting to shivering. Newly born human infants cannot shiver properly until they reach the age of one year. Consequently they possess brown fat between the shoulder blades, around the neck, heart, great vessels and lungs, to protect these vital organs from excessive cold.

Other important functions of fats, or more correctly, their constituent fatty acids, include their roles as integral parts of cell membranes and the myelin sheaths of nerves. The latter tissue facilitates high conduction velocities in vitally important nerve cells. Fats also provide excellent insulation against cold, as in the thick blubber layers in whales, seals and various Arctic mammals. The fatty insulation in the soles of the feet of the Arctic Fox have a low melting point (like soft margarine) so that the foot remains supple when the fox is standing or running on ice.

Interesting differences in the deposition of fat in men and women occur. The female sex hormone, oestrogen, plays an important part in this regard and is largely responsible for the more uniform deposition of fat beneath the skin of women. Before the advent of wet suits, the Japanese pearl divers were almost all women as the men could not endure the cold nearly as well. Why should men have a less uniform deposition of fat? No convincing answer is available but perhaps endurance running when hunting required that body heat be lost rapidly; a uniformly deposited fat layer would certainly retard heat flow.

Both fats and oils are used to increase the buoyancy of many aquatic animals, the oily livers of sharks being a good example. In some marine invertebrates, the larvae float about for several days as they slowly use up the energy in the oil droplets within their tissues. When the oil droplets have been used up, the larvae lose their buoyancy and settle on the bottom to start a new life. In this way they are dispersed uniformly within the habitat.

The wide-ranging functions of fats and oils and their almost universal distribution in biological systems makes them an ideal study for demonstrating both the unity and variety between life forms. This approach has been used to excellent effect by Neil Hadley in his lucid treatise on *The Adaptive Role of Lipids in Biological Systems*.

Secondary compounds

In the preceding pages you may have gained the impression that animals have completely free access to the nutrients and energy contained in plants. This is not the case and plants have evolved many protective mechanisms to defend themselves against herbivory, like tough outer coverings, thorns and resident insects; but most defend themselves using various forms of chemicals. These chemicals or secondary compounds, as they are sometimes called, are the result of millions of years of evolutionary war between plants and animals. This competition has sometimes been referred to as an arms race and judging by the masses of green material dominating most temperate landscapes, the plants seem to be winning.

The most common chemicals used for defense against herbivory are bitter tasting tannins. Green fruits often contain tannins to discourage their premature dispersion before the seeds contained therein are fully mature. When they ripen they usually turn an attractive colour; the tannins disappear and the sugar content rises dramatically. All these changes are designed to encourage animals to eat the fruit and disperse the seeds. This is perhaps one of the reasons why fruit-eating birds and primates are not colour blind.

Other important chemicals used for chemical defense among plants are alkaloids and so-called cyanogenic glycosides. The latter liberate poisonous hydrogen cyanide when the plant is damaged, which occurs when it is eaten. Well-known plant alkaloids exploited by humans are nicotine, strychnine and opium. The ancient Greeks used an extract of hemlock, rich in alkaloids, to "murder" the famous philosopher Socrates. In contrast, male danaid butterflies use plant alkaloids from *Senecio* plants to manufacture a perfume or pheromone, which they store in their winghair pencils; they eventually use this perfume to sexually stimulate the female butterfly during courtship. If the males are not successful in obtaining a dietary source of this alkaloid, courtship is not successful.

Cyanogenic glycosides occur in almonds, the kernels of cherries, apricots and apple seeds. Humans can become fatally ill from eating too many apple seeds, a fact which was understandably omitted from the story of Johnny Appleseed. Monarch butterflies exploit the glycoside toxins in milkweed plants by storing the cardiac glycosides in a harmless form within their tissues. This is then released when the butterflies are eaten by predators, which soon learn to avoid the monarchs by recognizing their characteristic warning coloration. This defense mechanism has become so effective that another species of butterfly, that contains no toxin in its tissues, has evolved the same warning colours as the monarch. Through this mimicry, it enjoys the same protection as the monarchs. However, there cannot be too many mimics as the signal to the predators would then become too weak.

Animals differ in their sensitivity to the secondary compounds used by plants for defense. It is an impressive sight to see a black rhino demolishing an euphorbia plant, with the highly irritating and toxic sap running over its lips. Just a drop of this milky sap on human skin causes very painful blisters. Also, many of the pungently flavoured herbs and spices we use in our various cuisines are part of the plant world's armamentarium. In fact, some people may even become addicted to the "hot" ingredient in red chillies and chilli peppers, known as capsaicin. This is now believed to stimulate the secretion of the morphine or opioid-like substances produced naturally within our bodies. John Prescott, of Australia's Commonwealth Scientific and Industrial Research Organisation, has shown experimentally that capsaicin increases taste sensitivity. He believes that food containing chillies seems to be more intensely flavoured for the devotee. Perhaps the combination of experiencing highly flavoured food and the sense of well-being induced by the opioid-like chemicals released into one's tissues are the reasons chillies and curries have become an essential ingredient in the cuisines of so many cultures of the world.

Similarly, it has recently been discovered that certain chemicals found in chocolate bind to the same receptor sites in the brain that interact with the active ingredient in marijuana. This may explain why "chocoholics" develop a desperate craving for chocolate and why so many of us find it such a soothing delicacy.

The procurement of energy from the environment by animals therefore makes a fascinating and complex study, which lies at the heart of zoology and ecology. From the simple and delicate absorption of nutrients across the cell membranes of unicellular animals, through the elegant cuisines of Japan to the crude mastication of elephants, countless variations in the manner of energy procurement have evolved. Nevertheless, the basic nutrients involved are very similar and, once they enter the complex chemistry of the cell, they become part of biochemical processes, which are for the most part shared by all animals. These processes probably date back, at least in part, to the primordial soup over three billion years ago. They also represent the essential cellular machinery for harnessing energy to keep entropy in check, at least temporarily.

Napoleon's
Water
Molecule

This chapter is somewhat of a digression from our theme of energy and animal life. It can, however, be justified by pointing out that energy exchange between living organisms and the environment would be impossible without water. Besides, the transport of energy-rich molecules from cell to cell in multi-cellular organisms is entirely dependent on water; or as Loren Eiseley would say, "If there is magic on this planet, it is contained in water."

We begin by examining some of the peculiar physical properties of water. This may seem a little daunting but this knowledge will allow us to appreciate some of the functions and fascinating journeys in which water molecules can be involved. Many of these peculiar properties of water are due to the weak bonds that the molecules form with one another. In addition to the strong chemical bonds which bind one oxygen atom with two hydrogen atoms to create the familiar molecule H_2O, some of the molecules are bonded together by rather weak forces known as hydrogen bonds, thus:

Structure of water molecules, showing the hydrogen bonding between hydrogen in one molecule and on an oxygen in a neighbouring molecule.

When water is frozen, nearly all the molecules will be hydrogen bonded; when it is a liquid many will still be bonded but far fewer, and when it becomes a vapour, none is bonded. At about 4°C water reaches its maximum density and, therefore, ice which is less dense at 0°C will float on the surface of cold lake water. The water molecules are packed more densely together at 4°C than in the solid form of ice. This peculiar physical property is of great significance in the cold regions of the world where lakes and ponds freeze on the surface, while the denser 4°C water collects at the bottom of the lake. Here various forms of life can escape freezing and survive a harsh winter. It is also fortunate that the colder water becomes, the greater is its capacity for dissolving oxygen.

Hydrogen bonding of water molecules is responsible for the strong surface tension on the surface of water; this is the reason why you can float a needle on water and why many insects, like gerrids and water striders, can walk upon water. Some send complex signals to one another by producing ripples of specific frequencies on the surface of the water. Because of our size we can plunge our hands into water with little or no effort, and even swim through it with reasonable ease. Think for a moment, however, about the energy that very small aquatic organisms have to use when moving through water, and especially when breaking the surface tension barrier between water and air.

The hydrogen bonding of water molecules is also responsible for the high heat capacity of water and the large amount of heat required to evaporate water into the gaseous or vapour form. The heat capacity of a substance is defined as the quantity of heat required to produce a given increase in temperature. For example, the heat capacity of water is about 3 000 times greater than air. This characteristic is most important in tempering the climate on our planet by damping wild swings in temperature, and in this regard it is significant that the 'earth' is mostly covered by water. It is also the reason we lose heat so rapidly when swimming in cold water.

The large amount of heat required to evaporate liquid water into the vapour form is important for the evaporative cooling of plants and animals. Most people have experienced the rapid cooling effect that alcohol, or especially ether, has when it evaporates from one's skin surface. With water the effect is slower but far more effective because sixty times as much heat is required to vapourize one gram of liquid water than is required to vapourize the same amount of ether. No wonder then that a great variety of animals as different as lions, ostriches, rhinos and even insects (e.g., the cicada) employ water for evaporative cooling. Some do so by sweating, others by panting.

The comparatively low freezing point and high boiling point of water is of great importance because, within the temperature conditions experienced over most our planet, water remains mostly in a liquid form, and life has largely evolved around liquid water. It is of course true that over millions of years the various ice ages have driven life, as we know it, back towards the lower latitudes. It is astonishing to think that only 30 000 years ago most of Canada was covered by an ice sheet several kilometres thick. Nevertheless "fluid life," built around the peculiar characteristics of water, re-established itself rapidly behind the retreating glaciers.

Water is a bipolar solvent and consequently it is an ideal medium for transporting nutrients, including energy-rich compounds, and a great variety of elements and molecules through plants and animals. The evolution, therefore, of all multi-cellular oganisms among both plants and animals was only possible because ever-present water provided a suitable medium for cells to communicate with one another.

When we refer to "animals" we usually mean vertebrate animals and forget that most of the animal kingdom does not possess a vertebral column and bony skeleton. Some invertebrate animals, especially aquatic ones, employ water as a skeleton within a flexible outer covering. Perhaps the extreme example of this would be an amoeba, which changes shape so markedly during locomotion that Linnaeus, who created the system for classifying plants and animals in the 18th century, named it *Chaos chaos*. The widespread evolution of hydroskeletons obviously took place because of the universal distribution of water, but also because water has a very low compressibility. Many marine creatures exploit this phenomenon in two ways. A hydroskeleton gives them support, but also considerable flexibility for changing shape. The high density of water, on the other hand, allows some animals, like squid, to gain momentum by ejecting water rapidly from their bodies, thereby employing efficient jet-propulsion.

The involvement of water in all the above situations is of crucial importance to life, but perhaps less fundamental than its role in assisting with the capture of energy from sunlight. Together with carbon dioxide, water provides the primary building blocks for the synthesis of simple sugars during the process of photosynthesis. Once these sugars are synthesized, chemical energy is available to the plant for countless biochemical reactions that are required for growth and reproduction. In this way the food chains are set in motion and a myriad of interactions between plants and animals become possible.

Having established the importance of the peculiar physical properties of water and reminded ourselves of its important functions, let us now amuse ourselves by following a water molecule on an imaginary journey around our planet.

We must first identify our molecule; it resides some 200 years ago in the leaves of a vine among the famous vineyards of the Médoc district in France. It exists as free water; it is not hydrogen bonded and is functioning as a solvent for sugars (energy) being transported from the leaves to the grapes which are forming on the vine. On arrival in a grape, the water molecule is exposed to various osmotic and hydrostatic forces which will determine its fate. The warm, dry summer air circulating around the vine will produce strong evaporative forces on the surface of the fruit and may draw the water molecule through the skin surface and into the atmosphere. But, the random collision of molecules within the tissues of one particular grape are such that our molecule remains within the fruit until it ripens completely. Thereafter the grapes are taken to the winery for pressing, controlled fermentation, filtering and clearing. Our molecule survives all these processes, even the tasting by the vintner.

Our molecule will spend the next three years in a storage vat; here it will continue to act as a solvent, becoming part of the secret chemistry of maturation as new fruity esters and several volatile compounds form to give the wine its special taste and bouquet. After three years the vintner bottles the wine and allows it to mature for a further five years. The particular bottle in which our water molecule came to reside is eventually commandeered by Napoleon Bonaparte's adjutant for provisioning the Emperor's personal supply of wine at the Russian front in 1812.

One can easily imagine how gruelling conditions at the Russian front must have been and how stressful for Napoleon to realize that all was lost, as his vermin and disease-ridden troops battled with hunger and the prospect of a bitter winter. Not surprisingly the Emperor turned to his favourite red wine for solace; and now we can follow the fate of our molecule once again as it is absorbed into his venous system, draining the blood back to his heart. From the heart the blood is pumped through the lungs, where again many rather random influences and collisions with other molecules will determine our molecule's fate. In this case fate determines that, as our molecule passes close to the very thin epithelium tissue lining the lung, it is exposed to a relatively low vapour pressure, and it follows other water molecules through the epithelium and into the air, filling the lung. Then, as the Emperor barks out his final decision to return to France, our molecule is expelled from his respiratory system.

Because of the cold night air it momentarily condenses with other water molecules on the Imperial breath, but soon turns again into unbonded water vapour as it is swept away on the cold wind.

Our unbonded water molecule can now be transported to any of the four corners of the globe, on thermals, updrafts over mountain ranges or coursing along the upper atmosphere in jet streams. Alternatively, it could become part of a large volume of ground water and be trapped in an underground vault for hundreds of thousands of years, becoming fossil water.

But let us rather imagine that our molecule, now in the form of water vapour, eventually arrives in a large air mass moving eastwards across the southern Atlantic Ocean towards the desert coast of Namibia, also known as the Skeleton Coast. In the direct path of this air mass is the very cold Benguella current which sweeps up the western seaboard of Namibia from the Antarctic. When the air mass moves over the cold current, many of the water molecules contained in this huge parcel of air lose much of their heat, the dew point is soon reached, and fine droplets start condensing together to form advective fog. This is then blown gently onto the extremely arid desert shore.

Several animal species make excellent use of the fog that condenses on the desert shore. One of these is the kelp fly. These flies complete their life cycles on kelp that has been stranded on the beach during storms and they use fog droplets, condensing on their wings, to assist them in maintaining a positive water balance. The flies use their legs to remove the condensed fog droplets from their wings and to transfer the droplets to their mouth parts. In this way our molecule has now become part of the body water of a kelp fly.

As the morning proceeds, the surfaces of the coastal dunes start to warm up and the fog begins to clear. When this happens an elegant little sand-diving lizard, *Aporosaura*, weighing only about two grams, emerges from beneath the sand to raise its body temperature by sun basking in a spread-eagled position on the sand surface. It warms rapidly to about 34°C and is then capable of extremely swift acrobatic manoeuvres that allow it to capture kelp flies, even when they are on the wing. I have found over sixty kelp flies in the engorged stomach of one of these small lizards, and it is clear that these lizards are enjoying two advantages by feeding on the flies. First, the flies are an excellent source of nutritional energy and, secondly, because the flies use fog water efficiently to maintain their own water balance, they are also an important source of water for the lizards.

Our water molecule has now moved from kelp fly to *Aporosaura* and, to complete the water chain, next to a snake, a side-winding adder which feeds upon these lizards and, eventually, to a chanting goshawk that perches on the crests of dunes in search of the well-camouflaged side-winders.

We could now send our molecule in any imaginary direction after it leaves the goshawk; it could, for example, become part of the photosynthetic process and be bound to a carbon dioxide molecule to form a sugar. The sugar could then be joined to other sugar molecules by the β-linkage to form cellulose and ultimately the woody portion of a tree.

Did you know that your wooden furniture and your house, if it is constructed of wood, consist mostly of sugar? The cellulose molecules formed in this way could possibly remain trapped in a valuable piece of antique furniture for hundreds of years; or, if the antique were to be burnt, the hydrogen atoms in the cellulose molecules would be stripped explosively from the mother molecule to combine with oxygen, thus forming entirely new water molecules.

Alternatively, our water molecule could find its way slowly to the Antarctic, where it might be incorporated into the algae living on the under surface of the pack ice. From here it would be an easy step to the shrimp-like creatures that feed upon the algae and thence to huge whales that strain the shrimp into their cavernous mouths. Our molecule could then be released into the atmosphere during the dramatic expulsion of spray around the blow hole of the whale, after it has migrated to the North Atlantic.

Then, imagine that a thirsty salmon swimming towards the Scottish coast is in danger of becoming dehydrated, because the surrounding sea water is far more salty than its own tissue fluids. Water, because of osmosis, therefore tends to leak out of the fish, particularly through the gills. To compensate for this, it drinks sea water, including our water molecule and then gets rid of the excess salts via specialized excretory cells in its gills. Eventually, the salmon reaches the mouth of its native river and moves from salt to fresh water; this poses an entirely different challenge. The fish is now faced with the danger of flooding, not dehydration, because the salt concentration in the fish's tissues is now greater than in the surrounding water and water begins to flood into its tissues. To overcome this, the blood supply to the salmon's kidney increases markedly and it pumps out the excess water in the form of large amounts of dilute urine. The latter, including our molecule, is soon further diluted by the river water and swept away on the current.

How big a jump will it now take to move our water molecule from a Scottish river to the glass of beer or cup of tea next to your elbow? Not a very big one, and when you next judge the colour of the wine in your glass, remember that one of the 5 000 000 000 000 000 000 000 water molecules in your glass may have passed through Napoleon's blood stream. Not impossible, if we recall Lord Kelvin's famous example, which involved marking all the water molecules in a glass of water by replacing the hydrogen atoms with deuterium atoms, and then pouring the contents of the glass into the ocean. After the marked molecules have been thoroughly mixed with the waters of all the oceans, one could remove a glass of water from any ocean and theoretically find about a hundred of the marked molecules in that sample.

There appears to be almost no end to the many functions that water can perform and the countless pathways it can travel. So much so that water has become synonymous with life and the flow of energy through living organisms. Consequently, many writers have been led to extol its physical properties as being ideal for life. This is, however, philosophical nonsense as water appeared on this planet long before life and the water molecule has been the core around which life has evolved and not the other way around.

The Thermostat
and
Why Elephants
and
Fleas Don't Sweat

This chapter is about the cost of keeping warm and cool, but first some important principles. Most animals are not able to maintain a constant body temperature by producing heat internally. Only birds and mammals are able to do so and they are called warm-blooded animals, or more correctly, endotherms. The rest of the animal kingdom, with very few exceptions, are cold-blooded and have to rely on a heat source outside the body to control their body temperatures. They are called ectothermic animals.

Interesting exceptions among the ectotherms are certain flying insects, which, as we have seen, shiver to raise the temperature of their flight muscles before take-off. Some mammals, like the naked mole rat that lives permanently in warm tunnels in the arid areas of Kenya, and is never exposed to extreme temperatures, have become ectothermic or cold-blooded. It takes on the temperature of the environment, much as a lizard might do. However, if lizards are provided with a heat source such as the sun (or a heat lamp), they will shuttle from shade to sun and, in so doing, maintain their body temperatures at a constant preferred level. This is known as behavioural thermoregulation and is probably best developed in giant monitor lizards, that will maintain a body temperature of 35° +/- 0.5° C indefinitely, given a permanent heat source. For this reason they are often regarded as occupying a position of transition between cold-blooded and warm-blooded animals. This may be so, but the python goes one better by using muscle contractions to produce heat when incubating its eggs. This is true physiological heat production and qualifies them as temporary endotherms.

Endothermy has many advantages. It allows endotherms to be independent of environmental heat and remain active at night or under cold conditions. Endothermy requires more energy and therefore an efficient oxygen supply to the tissues. This, in turn, has provided endotherms with a far greater capacity for sustained activity with all its many attendant advantages, particularly those involved in foraging and the capture of prey. There is, however, a price to pay and the energetic cost of free existence in a small mammal can be as much as thirteen times greater than in a similarly sized reptile. By far the greatest number of organisms on this planet are cold-blooded or ectotherms. Ectothermy must therefore have many advantages, even if it means not being able to enjoy the reading of Shakespeare's sonnets.

The energetic cost of keeping warm in endotherms depends mainly on the size of the animal, the insulation provided by its fur coat or plumage and the degree of cold to which it is exposed. The actual cost involved depends on the amount of energy required for shivering; this is a form of rapid muscle contraction, which does no work but produces heat. A very small mammal, because it loses heat rapidly, pays a high price for endothermy; it may begin to shiver at an air temperature as high as 28°C and at 10°C its metabolic rate may have increased two fold. Naked humans will on average start to increase their heat production when air temperatures fall below 26°C and at 5°C will increase their metabolic rate more than three times – obviously a costly process. In contrast, the white Arctic fox and several species of large Arctic mammals will only start to raise their metabolic rates at temperatures as low as -40°C. Therefore, if you are not well adapted, the cost of keeping warm can be high.

The cost of keeping cool is lower. In birds and certain mammals that pant to produce evaporative cooling, the muscles of the rib cage contract and relax rapidly to provide the mechanical pumping action required for panting. This requires energy, but as the muscles are contracting and relaxing at their natural frequency of oscillation, they are exploiting energy from the elastic recoil of the respiratory system. In this way appreciable amounts of energy are saved. Sweating, the other major mechanism used for evaporative cooling, is a glandular secretion requiring little energy. The cost of losing precious water due to panting and sweating can, however, be very expensive in desert conditions.

Let us now visit some polar and desert regions to find a few interesting examples to demonstrate the preceding principles.

Our first visit takes us to the forest/tundra border in northern Canada and while cross-country skiing on our first day out, we may disturb an Arctic ptarmigan that is using a snow drift as shelter from the wind. It is camouflaged to match the snow and, should we examine it, we will find that the plumage weighs eight times more than that of a similarly sized tropical bird. Moreover, the temperature in a sheltered roost 25 cm below the surface of the snow can be as much as 20°C warmer than the air temperature.

Other large animals in the area are also well insulated against the cold. In fact, as the size of the animals in the Arctic increases so does the length of their fur, with the musk-ox having the longest. The coat of musk-oxen consists of a long, outer protective coat that reaches almost to the hooves, and a woolly, thick undercoat which is shed in summer. Wool produced from this undercoat is reputed to be superior to the best cashmere. The annual cycle of growth and shedding of this coat is controlled by the seasonal change in the length of day, known as the photoperiod; this alters the function of the pituitary gland. Adult musk-oxen are large (the bulls weigh about 400 kg) and, because large size means a smaller relative surface area exposed to the cold, these animals are well adapted to withstanding the cold.

One might well ask how the small calves (8 kg) survive temperatures as low as -35°C shortly after birth. This represents a temperature gradient of about 73°C between the environmental and body temperatures and, as the temperature gradient is one of the most important physical factors driving heat loss, the young calves face the real danger of dying from exposure. Fortunately, the fur of the new-born calf is very well developed and extends over the legs, thereby affording excellent insulation, providing it remains dry. In addition, the new-born calves possess extensive deposits of the specialized fatty tissue known as brown fat, or more correctly, brown adipose tissue (BAT). Rapid oxidization of fats in BAT increases the metabolic rate of the newly born calf to a value almost thirteen times greater than the human basal metabolic rate, without the necessity of shivering. This phenomenon is known as non-shivering thermogenesis. Shivering, of course, also produces heat, but far less efficiently than BAT does, because it goes hand in hand with increased blood flow towards the body surface which, in turn, increases heat loss.

Far more common than the musk-oxen in this region are the caribou. They are also well insulated with a layer of fat and a thick hair coat. Their digestive systems are well adapted to feeding on the many species of lichens of the tundra. But what attracts our attention on the edge of the forest, between two big snowdrifts, is a large caribou, an aging male, that is attempting to gallop through the deep snow as it is pursued by a small pack of four wolves. The wolves are lighter and do not sink as deeply into the snow as the struggling caribou, but both the hunters and the hunted are now beginning to overheat. Vigorous exercise within a well insulated fur coat results in a marked rise in body temperature, in spite of the extremely cold environment. The caribou is panting deeply to pay back the oxygen debt it has incurred from the exercise. This rapid panting also serves another function of crucial importance, namely to keep the brain cool while the rest of the body heats up. This is accomplished by a remarkable heat exchanger, at the base of the caribou's brain, that any engineer would be proud to have designed.

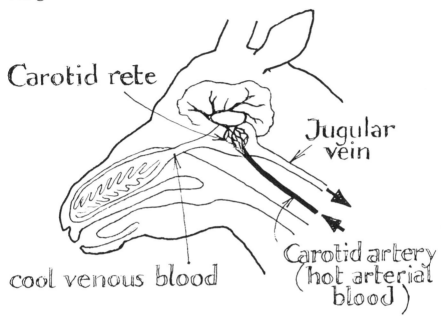

The antelope has a "radiator" in the nasal sinus and a heat exchanger for the brain.

This is how it works: because of the rapid panting, the membranes lining the large nasal sinuses of the caribou are cooled through the evaporation of moisture on their surfaces. This means that the venous blood draining from these sinuses also becomes cooled and, when it passes beneath the brain on its way back to the heart, it flows very close to a complex network of small arteries, carrying hot arterial blood from the heart to the brain. This arterial network at the base of the brain is known as the carotid rete and, because the hot arterial blood in the rete is flowing in the opposite direction to the cool venous blood, heat is efficiently transported from the arterial blood into the venous blood. The mechanism is similar to the heat shunt described previously in the muscles of the tuna and is also called counter-current heat exchange.

The arterial blood that eventually reaches the brain is therefore significantly cooler than it was prior to passing through the network or rete. In this way the caribou's brain is kept reasonably cool, in spite of an elevated body temperature. Unfortunately, for the caribou at least, the wolves possess a similar counter-current heat exchanger at the base of their brains. Eventually the caribou becomes exhausted and turns to face the wolves, but they decline this opportunity to attack a male still in fair condition. Perhaps they realize innately that an injured wolf has very little chance of survival in this harsh environment and that their pack is on the small side.

Another good example of overheating in sprinting mammals is provided by the swiftest of them all, the cheetah. Dick Taylor of Harvard has calculated that after a sprint of only 30 seconds, a cheetah's body temperature can rise by as much as 4°C to reach 41°C. This is equivalent to a serious fever in a human; but is well tolerated by the cheetah, largely because of the cooling action of the carotid rete on its brain. Fortunately for its major prey item in East Africa, the Thompson's Gazelle, the gazelle can sustain an even greater rise in body temperature when fleeing from a cheetah.

As we ski back to our forest camp, blessing our modern protective clothing, which provides excellent insulation and allows the loss of water vapour to keep us from overheating, we may begin to wonder what heat and cold are, in physical terms, and what the major physical factors are that affect their movement through various materials.

Heat is an expression we use to describe the rate of vibration of atoms and molecules which make up the matter surrounding us. This rate of vibration is usually referred to as kinetic energy or the energy of movement, and any matter above -273°C (absolute zero) possesses kinetic energy or heat. Temperature, on the other hand, is a man-made concept which is used to measure the mean kinetic energy within a prescribed system and the higher the temperature, the greater the amount of kinetic energy. There is really no such thing as cold, although it would be difficult to persuade an Inuk living within the Arctic circle that this was true. What we perceive as cold is merely a lack of kinetic energy, which then leads us to ask what the major physical factors are that influence heat flow through a body. They turn out to be the difference in temperature or temperature gradient between the body and the environment, the surface areas involved, the conductivity of the materials involved and the wind speed. In other words a small mouse with a thin fur coat in a deep freeze, close to a strong fan, will lose heat rapidly; whereas an elephant on a calm summer's day would hardly lose any heat. Wind speed can, through convection, increase either heat loss or heat gain, depending on the air temperature. Bedouins in the Sahara Desert therefore avoid hot winds as diligently as Inuit avoid cold winds in the Arctic.

Armed with this knowledge, let us now examine how natural selection has produced an animal in which heat loss has been reduced to a minimum, by moving farther north to the edge of the Arctic Ocean to do some seal watching. First of all, seals are large animals with a blocky, as opposed to a slender, shape; their surface area to volume ratio is therefore reduced, resulting in a slow rate of heat loss. In addition, there is a thick layer of blubber beneath the skin, which has a low conductivity. The blood supply to the blubber layer is minimal, creating a "cold shell" of blubber around the animal. This, in turn, reduces the temperature gradient between the skin surface and the environment and contributes to a general reduction in heat loss. Perhaps of greatest importance, however, is the seal's dense fur coat that traps a layer of air close to the skin surface. The air layer has such a low conductivity that it acts as an additional barrier to heat loss from the body. This is the major principle used in all cold-weather clothing and is easy to understand when we learn that the conductivity of water is about 25 times that of air.

When the seals enter the icy Arctic water their fur traps a layer of stagnant water close to the skin and heat loss is reduced by the so-called "wet-suit" effect. The stagnant layer of water is warmed, forming a boundary layer that reduces the temperature gradient between the animal and the surrounding water. In some marine mammals the undercoat is water repellent and a layer of air is trapped against the skin surface, which is even more efficient in reducing heat loss. In fact seals, like the caribou plunging through the deep snow, would overheat while swimming vigorously in these cold waters, were it not for the thermal windows provided by its naked flippers. As the seal's body temperature begins to rise, the blood vessels supplying blood to the flippers dilate and heat is rapidly lost to the icy surrounding water from the hot arterial blood coursing through the flippers.

Having examined the effects of relative surface areas, temperature gradients and conductivities of materials, all that remains is to examine wind speed to complete our elementary knowledge of the physics of heat transfer. Fortunately, the wind has died down completely in our Arctic camp and we can take off our shirts and enjoy the rather weak but welcome warmth of the sun, in spite of near-zero temperatures. How is this possible? The answer is to be found largely in the boundary layer or envelope of air which builds up on our skin surface. This thin boundary layer has a temperature significantly higher than the mean air temperature and, because of this, the temperature gradient between our bodies and the air immediately surrounding us is reduced and, consequently, the rate of heat loss as well.

The slightest increase in wind speed will, however, cause forced convective heat loss and see us scurrying for our clothing, as it only takes the slightest breeze to disrupt this thin layer of air molecules surrounding our skin. This is also one of the reasons why the wind chill effect is *relatively* greater at low wind speeds than at higher speeds. The natural "lie" of animal fur is disturbed at fairly low wind speeds, thereby disrupting the air layer trapped beneath the fur, resulting in more rapid heat loss. As a rough estimate, we can assume that heat loss from a lightly clothed person under cold conditions is proportional to the square root of the wind speed. Therefore at 16 km per hour heat loss is four times greater than in still air, whereas at 100 km per hour only 10 times greater. To be more accurate, wind chill effect is a combined measure of the effects of air temperature and wind on cooling and evaporation. For example, a human exposed to -40°C and a 50 km per hour wind would lose heat as if exposed to -80°C in still air.

Before leaving the Arctic we may be fortunate enough to see the eerie effect of solar particles on the earth's magneto-sphere, known as the Northern Lights or aurora borealis. If not, we can be reasonably certain of making contact with the giants of the Arctic, the polar bears. Their survival in this environment, which we humans interpret as harsh, is facilitated by their large size, their relatively small surface area, a layer of blubber beneath the skin and a thick pelage or hair coat. Why, however, are they white? Perhaps to camouflage them when they stalk their prey in the snow and on ice floes. But perhaps also because each hollow white hair may act as an optic fibre and reflect short wave radiation from the sun through the fur coat to the surface of the skin, which is darkly pigmented. This is a real possibility, but we must also remember that the colour of an animal has little or no effect on the absorption of heat if the wind speed exceeds about 4-5 metres per second – little more than a moderate breeze. This is because convection becomes far more impor-tant above this wind speed than the relatively small effect of colour. We could perhaps conclude that the need for camouflage in the snow is probably the main reason for the white colour of polar bears, but who is going to complain about a few fringe benefits?

A favourite trick question among zoology professors is to ask students how many penguins a polar bear should eat each day to satisfy its energy requirements. A polar bear, of course, would first have to migrate from the north to the south pole to taste this new delicacy, as penguins are restricted to the southern hemisphere. They are most un-usual birds and the contrast between their clumsy waddling on land and their elegant underwater swimming, reminiscent at times of a torpedo, seldom fails to amaze and enchant spectators. Perhaps the most interesting species is the Antarctic emperor penguin, that has astonishingly chosen mid-winter as its breeding season. Moreover, both male and female penguins trek as far as 100 km inland from their normal marine habitat to reach their breeding site.

On arrival at the breeding site, the female lays a single egg, that the male bird incubates by holding it between its feet and a fold of the abdomen which partially envelopes the egg. During incubation the male survives the intense cold of the Antarctic winter (-30° to -40°C), not to speak of the wind chill effect from the strong Antarctic winds. But a major expenditure of energy is involved, as the males lose up to 40 percent of their body weight during this period of starvation. They reduce the effect of the cold and the chilling effect of the wind by huddling together because this reduces the surface area exposed to the cold air. If a male bird is prevented from huddling in experiments, it uses twice the amount of energy compared to when it is allowed to huddle naturally.

The female abandons her freshly laid egg to return to the shore line, where she feeds intensively to restore her fat reserves. After about two months she will return to the rookery to feed her chick on regurgitated stomach contents and to relieve the male from his paternal duties, so that he can return to the ocean. We can only guess how this elaborate breeding behaviour evolved and what selection pressures drove these animals to such an unusual breeding pattern. Was predation too intense at the shore line? To what extent did their physiology, particularly their ability to lay down large amounts of fat, pre-adapt them to this breeding pattern? An interesting riddle for ornithologists.

Humans cannot compete with penguins in this regard and have never lived permanently in Antarctica. But some 27 000 years ago when the last ice age had entered its coldest phase and the sea level was 100 metres lower than today, human hunters were roaming the tundra and steppe. Eight thousand years ago hunters were well established within the Arctic circle. It is in fact remarkable how adaptable humans are to extremes in climate. This is largely because we possess a naked skin and can markedly alter our conductance by selecting suitable clothing. Our intellectual ability to plan ahead and build appropriate shelters is also important.

Nevertheless, we still do not understand how humans become adapted to the cold. Men working on oil rigs in the far north exhibit several signs of improved adaptation to cold after some six weeks of exposure to the climate. Most noticeable is their superior manual dexterity under cold conditions, when compared with new arrivals at the rig. Physiologists believe that the blood vessels supplying blood to the fingers dilate at intervals in cold-adapted workers but remain constricted in the new arrivals.

Could cold-adapted humans produce heat by non-shivering thermogenesis, as the young musk-ox calf does, by breaking down fatty acids far more rapidly than is possible in ordinary white fat? Although newly-born humans exhibit non-shivering thermogenesis, there is very little reliable evidence to suggest that adults are able to exploit this mechanism. The late P.F. Scholander of the Scripps Institute in California believed that Alacaluf Indians, inhabiting the southern tip of Tierra del Fuego, had evolved specialized adaptations to cold. These natives were made famous by Darwin when he voyaged around Cape Horn in the Beagle and described their almost miraculous resistance to cold. Apparently they bathed in ice water and walked barefoot through snow-fields with an absolute minimum of clothing. Scholander found that the resting metabolic rate of the Alacalufs was consistently higher than expected, even when they were not shivering. It is possible that they were using non-shivering thermogenesis, but we shall never know for sure as the Alacalufs have died out as a separate group.

Why would non-shivering thermogenesis be an advantage over shivering? Because, as we have learnt, muscular activity during shivering leads to increased blood flow to the surface and consequently increased heat loss. Nevertheless, it is interesting to note that the percentage of heat stored during shivering is still as high as 50 percent, whereas when exercise is used to generate heat to overcome cold, only about 20 percent of the heat produced is stored.

Humans cannot, however, rely on adaptation when faced with extreme cold. Hypothermia is a very dangerous condition and the newly born and the aged are particularly susceptible; the former because of their high relative surface area and inability to shiver efficiently. The aged respond less sensitively to cold stimuli and this is often aggravated by poor socio-economic conditions. What happens to us when we succumb to hypothermia? The first response is naturally intense shivering, but as our body temperature approaches 30-32°C (37°C being normal), the intensity of shivering starts to wane and our mental acuity declines to such an extent that we begin to make irrational decisions. If lost in the snow, we might start walking in circles. Between 30-32°C shivering will cease and most people will lose consciousness. At body temperatures of 25-28°C the heart begins to beat irregularly, blood pressure falls and the ventricle of the heart starts to fibrillate (irregular rapid twitchings of the muscular wall of the heart). Most people will succumb to heart failure at body temperatures of around 18-22°C.

The treatment of people suffering from hypothermia has become a highly specialized discipline and beyond our scope, except to remark that every effort should be made to avoid ventricular fibrillation which precedes heart failure. The classic rescue of hypothermic patients in the Swiss Alps by a Saint Bernard dog, with a brandy cask around its neck, has alas lost the approval of the medical pundits. Instead, the pharmacologists will soon provide us with a tailor-made molecule which will produce body heat without any pleasant side-effects!

More challenging is the practice of g Tum-mo yoga, that has been recommended by some physicians to increase the production of body heat. This practice is essentially a form of Tibetan Buddhist meditation. Professor Herbert Benson and his colleagues from Boston travelled to the Himalayas to study this practice and found that their subjects could increase the skin temperatures of their toes and fingers by as much as 8.3°C during meditation. Apparently novices, learning the art of this type of meditation, practice by wrapping themselves in sheets dipped in icy water. The novice must then dry the sheet by using his body heat and, as soon as the sheet is dry, it is again immersed in ice water and wrapped around him.

This procedure can last from sunset until day-break and the person who dries the greatest number of sheets is declared the winner of the competition. This remarkable feat is most likely achieved through voluntary dilation of the small arteries near the surface of the body and it has been calculated that even during simple transcendental meditation, blood flow through the kidneys and liver is decreased by as much as 44 percent. Presumably the blood flow is diverted to either the brain or skin surface or both.

By way of contrast let us now move from the Himalayas to the Sahara Desert, where we can examine the behaviour of a species of desert ant, *Cataglyphis bombycina*, that survives in the desert by walking a thermal tightrope. Swiss and South African zoologists have studied this species in detail and found that it feeds primarily on insects and spiders that have recently succumbed to heat death on the extremely hot sand surface of the Sahara. In order to maximize the probability of encountering such prey, the ants forage at very high surface temperatures, which soon become too hot and they are then obliged to return to their nest. This means that the ants employ a very narrow thermal window, so to speak, for foraging. The foragers erupt from the nest in the form of a mass exodus as soon as a threshold temperature of 46°C is reached. After approximately 10-15 minutes above the surface, their body temperatures begin to approach the critical maximum of 53°C and they all return to the nest to escape thermal death. This rapid foraging pattern not only increases the probability of the ants encountering their disabled prey, but it also allows them to avoid their own major predator – a species of Saharan lizard; and herein lies a simple but significant lesson in the physics of heat transfer.

Simple geometry again tells us that the tiny ants have a large surface area relative to their volume and this would at first appear to be a great disadvantage, as it means that they will gain heat very rapidly from the hot Saharan environment. But it also means that they are able to off-load heat rapidly and, by interrupting their rapid running across the desert surface at regular intervals to climb onto a dried grass stalk or twig, they are able to shed excess heat rapidly to the slightly cooler environment a few centimetres above the surface. Although the lizard has the advantage of a smaller relative surface area, its larger size confers a greater degree of thermal inertia upon it and consequently it cannot off-load heat nearly as rapidly as the ants do. The lizards therefore retire to a cooler refuge before most of the ants exit from the nest.

One must presume that the temperature sensing of the ants is not always perfect and that they must on occasion leave the nest a little too early. Otherwise the predator/prey relationship between the lizard and the ants would not survive. In any event, *Cataglyphis* seems to be pushing its luck to the very limits of survival in order to occupy a vacant niche.

Thermal inertia is not always a disadvantage in the desert and for this reason large animals that heat up slowly are often very well adapted to desert life. One of the great pioneers in environmental physiology, Knut Schmidt-Nielsen of Duke University, has shown this in his studies on Saharan camels. When camels are deprived of water, which frequently happens in the desert, their body temperatures fluctuate from around 34°C at daybreak to reach about 40°C during the peak heat of the day. Bearing in mind how big camels are, this represents a storage of a large amount of heat; and the great advantage of this form of adaptation is that significant amounts of precious water are saved, which would otherwise be lost in evaporative cooling to maintain a constant body temperature. The stored heat is then lost to the cool night air by radiation and convection, without employing sweating or panting – a great advantage in desert conditions.

The oryx or gemsbok is another large antelope of the desert that employs heat storage as a means of saving water. Few sights can compete with the thrill of watching these splendid antelope, with their long unicorn-like horns, galloping at full speed between desert dunes. (With a little imagination one is also reminded of the caribou floundering through snow drifts pursued by wolves.) Here the oryx are being pursued by a pack of hyenas, but both the oryx and the caribou are employing the same kind of heat exchanger, the carotid rete, to keep their brains reasonably cool while their deep body temperatures rise dramatically from the vigorous exercise. Usually the race will not last too long. As soon as the hyenas start to tire and overheat, they will abandon the chase. Like the wolves they prefer to tackle the weak and infirm.

This is not the case when humans use helicopters and trucks to chase and capture antelope. With two of my students, Ian and Retha Hofmeyr, I found that under these conditions the animals are often chased beyond their natural limits of endurance; they develop a large oxygen debt and overheat, reaching fever levels. Because insufficient oxygen is reaching the muscle cells, glucose is no longer fully oxidized to carbon dioxide and water to provide energy. Instead, glucose is catabolized to lactic acid, which provides less energy, but does not require oxygen. Eventually, the large amounts of lactic acid accumulating in the muscles cause an irreversible muscle disease known as capture myopathy. The animals die within several days in great pain or, mercifully, sufficient potassium leaks from the muscles to cause heart failure. Modern game capturing methods avoid these problems by herding the animals much more slowly and by using suitable tranquillizers. In this regard it is significant that race horses, which have been artificially selected and trained for endurance running at high speed, have far superior endurance compared to antelope and their wild cousins, the zebras.

While discussing large polar animals, we noted that the larger the animals became, the thicker their fur or insulation became. This is understandable as all polar animals would benefit from the thickest possible insulation, but there is a limit to the length of the fur that a small animal can carry without interfering with its locomotion.

The opposite situation holds true for the wonderful array of antelope inhabiting the African savannah. Only the smallest antelope such as the steenbok and klipspringer have any appreciable insulation in the form of a fur coat. The larger the antelope become, the thinner the hair coat becomes. In large antelope such as kudu and oryx it is only a few millimetres thick; while the very large mammals, the rhinos and elephants, have no fur at all. We can interpret this as an adaptation to facilitate the loss of body heat, because as the animals become larger, their relative surface area becomes smaller, making it increasingly difficult to lose heat. This is a tough compromise to make because the African savannah also experiences cold weather at times and rapid heat loss through a thin hair coat results in long periods of shivering. This uses a lot of energy and reduces the potential of these animals to store reserve calories in the form of fat. Therefore, fat stores in African antelope are often limited and a cold spell after prolonged drought can lead to large scale mortality. It is also the reason, in more mundane terms, why chefs have to lard African venison before roasting it, because of its very low fat content.

In hot deserts, most of the animals are small ectotherms, such as spiders, scorpions, insects and small reptiles. They can hide readily from the extremes in climate and regulate their temperatures behaviourally. Some are so adept at avoiding extreme heat, cold and aridity, that they do not in fact "live" in the desert, but rather in a small favourable microclimate. There are only a few species of large and middle-sized mammals in the desert. The small to medium-sized mammals generally live in underground burrows and only emerge when conditions are favourable.

How do humans cope with excessive heat? We know more about this than our adaptation to cold. It would seem that the major adaptation to heat in humans is an increased ability to sweat. Blood volume also increases so that when the superficial blood vessels dilate in response to a warm environment, there is no sudden decrease in the blood pressure, leading to fainting. An increased sweating rate is the most important adaptation and exposure to heat for a few hours per day will be sufficient to increase the sweating rate in everyone, irrespective of age, sex or race. Within a month, a daily visit to the sauna will ensure that a Scotsman from the outer Hebrides will be as well adapted as a Bedouin to the Sahara Desert's heat, providing he dons the Bedouin's robes to protect him from direct solar radiation.

The importance of sweating in humans may be related to our lifestyle in our recent past. As hunter/gatherers and scavengers there must have been many occasions when endurance running would have been a major advantage. Unlike the antelope, dogs and big cats, we do not possess a carotid rete to cool our brains and therefore rely on keeping the whole body cool by sweating actively. Together with the horse family, we are the champion sweaters of the whole animal kingdom and perhaps our nakedness has developed to enhance the cooling effect of perspiring. Also, unlike dogs, cats and many antelope, we do not respond to excessive heat by employing thermal panting – perhaps because of our dependence on complex verbal communication.

To return to the carotid rete, it would seem that possession of a rete is limited to those mammals with a fur coat that have to sprint away from swift predators. The predators in turn also have fur coats and possess a carotid rete. Wild African pigs are an interesting exception; they do not have a fur coat but do have to sprint from predators on occasion and, as they do not have functional sweat glands, they require a carotid rete to cool their brains. The English expression "to sweat like a pig" is therefore nonsense.

Rhinos sweat freely and this is perhaps why their ears are so small compared to an elephant's. Elephants, because of their relatively small surface area, low metabolic rate and their great thermal inertia, do not have to sweat actively. In fact, their skins do not possess sweat glands. Nevertheless, studies at the University of the Witwatersrand by professors Wright and Luck showed that they do lose water vapour through the skin at an average rate of 0.23 litres of water per square metre per hour. This will have some cooling effect, but by far the main window of heat loss is through the greatly enlarged ears, which account for 20 percent of an elephant's surface area. When these animals become overheated, they increase the flow rate of hot arterial blood through their ears to as much as 12 litres per minute. Heat is then lost through the relatively thin skin of the ears by radiation and convection. Fanning of the ears while standing in the shade will naturally facilitate convective heat loss.

Is there an evolutionary pattern with regard to animals which rely on sweating and those which rely on panting to get rid of excess heat? Not really. A large oryx antelope both sweats and pants if it is given sufficient water, whereas the wildebeest or gnu, with its huge head and prominent Roman nose, only employs thermal panting. What is certain, however, is that very small animals neither sweat nor pant as they would lose too great a percentage of their body water and desiccate completely. For example, a camel uses only one percent of its body weight per hour for evaporative cooling under hot desert conditions. In contrast, a 100 gram desert rat would use about 13 percent of its body weight per hour under similar conditions and face fatal dehydration within two hours. Needless to say fleas, were they to sweat, would desiccate within minutes because of their small size and do not even "consider" sweating!

Several reasons have been advanced to explain why the pro-hominids developed an upright or two-legged stance during human evolution. The ability to look over tall grass in the savannah to avoid predation is a possible reason; so is the obvious advantage of the freeing of the hands to use tools. The ability to carry home scavenged meat and other foods is yet another well-known reason, but perhaps the most imaginative reason to be advanced to date is that described by Pete Wheeler of Liverpool. He has pointed out that an upright stance would reduce the profile of these primates exposed to direct solar radiation, and at the same time lift the head into a much cooler environment. Anyone who has experienced the heat near the surface of the soil in the African savannah would agree that this would have been a distinct advantage, especially when we recall that primates do not possess a carotid rete.

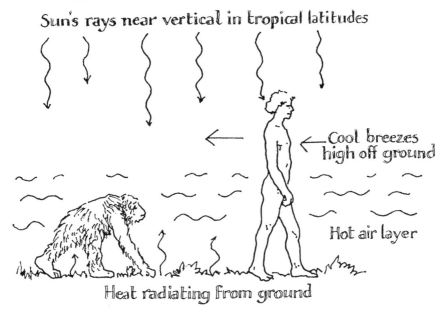

Stand tall and stay cool! Bipeds live in a cooler environment than quadrupeds.

Modern humans are among the most precise thermoregulators in the animal kingdom. Next time you are among a group of twenty people or so, look around and you will probably find someone who can tune a violin very precisely, someone who can draw rather accurately, someone who can solve a calculus problem or someone who can write a few lines of poetry. If you then pass a thermometer around the room you will find that they all have a remarkably uniform body temperature. Surely it is no accident that our intellectual development has gone hand-in-hand with the evolution of very precise temperature regulation of our central nervous system. When our brain temperature drops to around 32°C our mental responses are severely impaired; when we are running a high fever, we frequently become delirious.

In contrast, fleas and elephants are not precise thermoregulators and do not sweat. But without sweating we may never have become philosophers!

Enough
Energy
for
Sex?

We are all so preoccupied with the immediacy of our everyday existence that we seldom question how our distant ancestors coped with life or, in fact, how many ancestors preceded us. If we go far enough back in evolution, our existence today once depended on the precarious survival of an amoeba-like creature hundreds of millions of years ago. If we then trace the number of generations that our DNA has passed through since then, to arrive at the end of the twentieth century, the number becomes overwhelming. We need not, however, go as far back in evolution as the arrival of the first unicellular animals. Let us rather consider the therapsids or mammal-like reptiles from which the mammals arose.

These reptiles lived some 190-280 million years ago and probably reproduced in typical reptilian fashion through internal fertilization followed by egg-laying. This means that each pair of therapsids that reproduced successfully had to obtain sufficient energy from the environment to reach sexual maturity. The males would then have to expend energy in finding a suitable mate and then perhaps fight off competing males before performing an elaborate courtship ritual to capture a female's fancy. The female would also have required sufficient energy to grow to maturity and then enough energy and protein to yolk her large eggs. In other words, all the people you see around you today owe their existence to countless successful matings over millions of years which, in turn, depended on each ancestral couple obtaining sufficient energy from the environment.

The above remarks also hold true for many plants that reproduce sexually, although courtship rituals are not necessary. Instead, plants invest energy in nectar, elaborate flowers, fragrances and enticing fruits for a similar purpose. One of the more interesting examples of this kind of energy investment is provided by the voodoo lily, *Sauromatum guttatum*. These plants are able to increase the temperature of their flowers briefly by as much as 22°C above the environmental temperature. When this happens the energy turnover in the inflorescence increases 100 fold and is comparable to the metabolic rate of a humming bird. What is the purpose of this burst of heat production? At these high temperatures putrid-smelling amines and indoles are liberated into the surrounding air, which attract pollinating insects. The chemical responsible for synchronizing the energy burst with the opening of the flower is salicylic acid, the active compound in aspirin.

We return to the animal world, and by using the reproductive cycle of red deer in the Scottish Highlands as an example, we can examine the cost and nature of reproduction in a large mammal. Thereafter, we shall conclude by asking the question of why sexual reproduction evolved.

Our visit coincides with autumn and the days have been shortening for several months. This gradual change in daylight-length has been perceived by the stags for many weeks and this message has been conveyed via the optic nerves to that part of the brain known as the hypothalamus. The hypothalamus lies just above the pituitary gland at the base of the brain, where it fulfils its role as leader of the endocrine orchestra, largely by controlling the release of hormones from the pituitary gland. In the present instance the hypothalamus, in response to diminishing daylength, has brought about the release of two pituitary hormones; these, in turn, have stimulated the production of sperm and the male hormone testosterone by the testes.

The testosterone has resulted in profound changes in the physiology and behaviour of the males. The antlers, which until now have been growing beneath their covering of soft "velvet," become hardened and the velvet is shed. Aggressive male behaviour is the order of the day with stags strutting, roaring and pawing the ground. If two stags compete for the favours of a female, the contest will be decided by locking antlers and some intense jousting. Usually the heaviest stag with its larger antlers will win the battle. This does not, however, mean that the winner will be automatically accepted by the doe. He must still expend considerable energy in harassing and cajoling her into mating.

This competition among stags is not unique. It is a widespread phenomenon in the animal kingdom from desert beetles to magnificent male lions. It is a behaviour pattern that can consume large amounts of energy, particularly in those animals like wildebeest, or gnus, that defend their parched territories against other males for long periods. This struggle between males for the possession of females was called sexual selection by Darwin and it is accepted as an important process of natural selection by biologists today. In this context, sexual selection is acting as a genetic filter to improve the probability that the "fittest" males will win the favours of the most females. This filtering process pays off as the reproductive process in males is not nearly as costly as in the females of most species, and a single victorious male can fertilize many females. The importance of natural selection through female choice is, however, still a contentious matter.

We should also remember that the females of many animal species do not choose only a single mate. In many species multiple mating is the rule rather than the exception. It has also been suggested that in certain female snakes, that are able to store sperm in their reproductive tracts for several months, multiple copulations allow the females to store the sperm of a variety of males. They then release all the sperm into the oviduct simultaneously to compete with one another to fertilize the ova. This argument presupposes that highly mobile vigorous sperm also carry the genes for high fitness in the offspring, which may not be the case.

Scientists have contended that large antlers confer a superior position on stags within their male hierarchy, and that those with the largest antlers most often exhibit the highest circulating levels of testosterone in their blood. There is also some evidence that if a set of large antlers is removed artificially from a stag and replaced with a small set, the stag's position in the pecking order drops dramatically and this is followed by a significant drop in testosterone production.

Meanwhile the diminishing daylength has also been influencing the pituitary glands of the females via the hypothalamus. Small follicles in the ovary have been stimulated to grow and these produce increasing amounts of the female sex hormone oestrogen until a peak is reached when the animals come into heat or, more correctly, into oestrus. The term oestrus is derived from the Greek *oistros*, meaning gadfly. These worry cattle to such an extent that they are driven into a frenzy. Consequently, the excitable behaviour of a female mammal in heat became known as oestrus. During oestrus the female is receptive to the male for a short period (several hours) and will only allow copulation with a male of her choice during this period. This is important because the escape of the ovum from the ovary (ovulation) into the female reproductive tract occurs towards the end of the oestrus period. In this way, insemination is closely synchronized with ovulation, thereby optimizing the possibility of fertilization.

Let us now assume that fertilization in the doe has been successful and then follow her pregnancy to the birth and suckling of her fawn. This will also allow us to learn how a female mammal virtually surrenders her hormonal or endocrine system to the foetus during gestation.

Once fertilization has occurred in the oviduct, the fertilized ovum makes its way to the uterus assisted by cilia-like hairs, within the oviduct, which beat in the direction of the uterus. Meanwhile, the follicle that released its ovum has been transformed into the so-called yellow body or corpus luteum. The latter now secretes the hormone of pregnancy or progesterone, which ensures that the ovum implants into the wall of the uterus and that the uterus remains quiescent during pregnancy. The tissue which forms at the interface between the embryo and the maternal circulation is known as the placenta. It secretes a hormone which promotes the survival of the corpus luteum and therefore adequate progesterone throughout pregnancy. This, then, is the first example of the embryo's control over the maternal endocrine system.

Pregnancy will then proceed to term. In the case of red deer this will take 25 weeks. What, however, triggers the end of pregnancy? It would be logical to have this trigger mechanism controlled by the foetus: it "knows" best when it is ready to enter the world, and this is in fact what happens. Once the hypothalamus in the brain of the foetus has matured, it causes the release of a hormone from the foetal pituitary, which stimulates the secretion of cortisol from the adrenal glands of the foetus. The cortisol is then responsible for several important hormonal changes taking place in the doe. For example, secretion of the female sex hormone oestrogen increases markedly, whereas progesterone declines and the muscles of the uterus wall become highly sensitive to stimuli to contract readily. This is the second example of foetal control over the maternal hormonal system. The third example is provided by the mechanical distension of the uterus and cervix by the sheer size of the foetus. This causes the release of a very important hormone from the maternal pituitary gland, known as oxytocin. The latter causes powerful contractions of the uterus and expulsion of the foetus.

The final example of foetal control occurs shortly after birth when the fawn begins to suckle. The doe's mammary gland has developed in size and complexity during the last several weeks of pregnancy, mostly under the influence of oestrogen and progesterone. Now, the actual suckling stimulus is conveyed to the brain via the spinal cord, and from there to the pituitary gland, where the release of a most interesting hormone known as prolactin occurs. Prolactin is then responsible for the actual secretion of milk by the well-developed mammary gland.

Prolactin is a hormone with an ancient lineage because the same hormone is responsible for stimulating the secretion of crop milk in birds, for controlling salt secretion in the gills of fishes and for stimulating the secretion of mucus on the skin of certain fish. In one particular fish species, the newly hatched young have been seen to feed off the skin surface of the parental fish. Since we know that mammary glands are modified skin glands, then we can speculate that prolactin was involved in the very first primitive form of lactation.

By definition, lactation is common to all mammals. Nevertheless, evolutionists still ponder on how and why lactation evolved, and their arguments are of interest. It is easy to understand how mammary glands could have evolved from existing skin glands such as sweat glands; the most primitive form of "lactation" was probably the secretion of a watery fluid by ventral skin glands to keep the eggs moist in an egg-laying mammal's nest. The why part of the question also has some obvious answers. Lactation allows a mother to deliver a large amount of food in a short space of time. This is of great advantage to mothers that have to leave their young in dens or nests while they spend long periods hunting or foraging. It may surprise you to learn that some marine mammals only suckle once per week. Long before the onset of lactation, a female can store nutrients in her body during a favourable period and then break those stores down to feed her young during lactation, even if unfavourable environmental conditions arise soon after the birth of her offspring.

Many mammalian herbivores are born in an environment which offers only coarse fibrous dry plant material as forage. They cannot use this material and require highly nutritious food at this critical stage of their development. The adult ruminant animal can, however, convert this fibrous material through fermentation in its capacious fore-stomach into nutrients that can readily be transformed into milk by the mammary glands. But perhaps one of the most important advantages of lactation is that it strengthens the bond between mothers and their young. This is especially important in those mammals that have to acquire learned behaviour for survival, as is the case in the primate, dog and cat families.

Before leaving lactation, we should not neglect the differences in milk composition that exists among mammals. Milk is a highly nutritious fluid containing vitamins, minerals, fat, sugar and a very high quality protein. In some mammals it can be deficient in iron but sucklings usually supplement this deficiency from another source fairly soon after birth. However, how do we explain some of the marked differences in milk composition among mammals? Cow's milk, for example, contains about 3.5% fat, whereas rhino milk has only 0.3% fat and the milk of blue whales over 53.2% fat. The diluted nature of rhino milk is probably to compensate for the evaporative water loss experienced by the rhino calf under the hot conditions characteristic of the African savannah. In the case of the blue whale, the calf suckles infrequently and must build up a thick layer of blubber rapidly to facilitate thermoregulation and provide buoyancy; hence its rich milk. The discoverer of vaccination for smallpox, Edward Jenner, described porpoise milk as "creamy rich" as early as 1773.

The description of the reproductive cycle in red deer should not leave the reader with the impression that all mammals exhibit similar cycles. For example, some mammals, such as rabbits, remain in a constant state of oestrus and only ovulate after the stimulation of copulation. They are known as induced ovulators and this pattern of ovulation results in an almost simultaneous arrival of the ova and sperm into the female tract.

In other species, the implantation of the fertilized ovum in the uterus is delayed until environmental conditions provide the required stimulus. This means that mating and fertilization do not have to predate the most favourable time of the year by the length of the normal gestation period, thereby making it easier to find a mate. Typical examples of animals that exhibit delayed implantation are roe deer, badgers, kangaroos, seals and sea lions.

The human cycle is also odd in that there is no overt period of oestrus; ovulation occurs "silently" and there are no safeguards to synchronize the arrival of sperm and the ovum in the oviduct. Perhaps frequency of coitus compensates for this theoretical disadvantage. The exploding human populations in the world would seem to confirm this view.

Red deer are also an example of seasonal breeders. As we have seen, the stags and does both respond to diminishing daylight-length in the autumn by becoming more sexually responsive; the courtship behaviour of the stags and the pheromones of the does in oestrus are responsible for the final stimulus for mating. Shortly after the rutting season, testosterone secretion in the stags begins to decline until it is too low to maintain the antlers; they then cast off their antlers. During the next summer, the stags grow new antlers within a period of about 12-16 weeks. This represents the deposition of about 3 kg of hard antler and the process places heavy demands on the animals' mineral resources. They therefore attempt to supplement their mineral intake by gnawing their cast antlers of the previous season, and by ingesting considerable amounts of mineral-rich soil.

The gestation period of the does ends in spring and the fawns are born at the most favourable time of the year. The flush of green vegetation in the spring also provides sufficient energy and protein for the heavy demands of lactation. The dependence of red deer on daylight length for synchronizing their seasonal reproductive cycle is not surprising, as it is the most dependable environmental variable, remaining exactly the same from year to year. Many other mammals also depend on daylight length, but in those in which the gestation period is very short and the young mature rapidly, they very often respond directly to a favourable environment. A good example of the latter would be a small desert rodent that can reproduce swiftly after rain has fallen in the desert and brought about a sudden improvement in the availability of food. Similarly, desert birds have been observed to start singing and nest-building shortly after water was poured down the front of their cages after a prolonged drought. In these instances it would not be appropriate for the species to depend on daylight length, as the arrival of favourable environmental conditions in the desert is not always associated with a specific period in the cycle of changing daylength.

Similarly, most desert plants respond directly to rainfall by growing and reproducing very rapidly. A good example of this is provided by the spectacular display of field after field of desert flowers after winter rains in the Namaqualand area of southern Africa. Many of the flowers that provide these massed displays belong to the daisy family and the flowers only open above a certain temperature threshold. Thereafter they track the sun all day by pointing their flowers, shaped like miniature dish antennae, towards the sun. In so doing they provide a micro-climate within the small flower that is significantly warmer than the surrounding air. Insects take shelter here and inadvertently fulfil the role of pollinators. Because these plants are ephemeral, the higher temperature of the inflorescence also speeds up the ripening of the seeds and shortens the life cycle – a distinct advantage for a desert plant with a compressed life cycle.

It would also be inappropriate for seasonal breeding to occur in those animals that have a highly structured social organization that depends, to some extent, on division of duties. African wild dogs do not exhibit seasonal breeding and the females that are without pups will help feed the pups of the breeding female. They do this by returning from the kill to regurgitate food for the youngsters. On the other hand, wildebeest or gnus have a very loose social structure and breed seasonally; the females belong to huge herds and usually calve within a few weeks of one another, thereby swamping the environment with calves. In this way many calves survive, as there are far too many for the predators to kill. In contrast, the close companions of the wildebeest on the African savannah, the plains zebras, do not breed seasonally as they rely on close cooperation of the females as sentinels within a tight family group. The higher primates are often not seasonal breeders for the same reason, humans being a good example. Just imagine the chaos if this were not so.

Many insect species are not strict seasonal breeders because they can respond rapidly to favourable environmental stimuli and because they usually have a short life cycle. One of the most interesting of these examples is provided by Miriam Rothschild's studies on rabbit fleas. Apparently the sexual cycle of the fleas, attached to the rabbit, becomes synchronized with the rabbit's hormonal cycle during the latter's short pregnancy. This synchronism is controlled by the rabbit's hormones contained in the regular blood meals which the fleas are taking from mother rabbit. When the rabbit's young are born, the female fleas have reached the peak of their sexual cycle with many ripe eggs. They then leave the mother rabbit to lay their fertile eggs on the young rabbits, thus ensuring the continued dispersion and survival of their species.

Seasonal breeding makes good sense as it is an insurance against the unnecessary wastage of energy. It is also easy to understand how it evolved: those females that breed outside the most favourable season would most likely lose their offspring and these aberrant genes would in time be removed from the gene pool. But what about the question of the overwhelming diversity of reproductive patterns that have evolved on this planet; why and how did they evolve?

Let us digress briefly to mention a few examples: certain starfish, such as *Asterias amurensis*, release around 100 million eggs, whereas the carpenter bee, *Xylocopa capitata,* may lay only four eggs in a lifetime. Some animals produce huge numbers of young and die shortly thereafter; others produce very few, and take great care of their young for long periods and reproduce repeatedly over a long life span. In addition to the diversity of reproductive patterns, courtship behaviour also varies tremendously. The scorpion's dance, the lamps of glow worms and fire flies, the calling of frogs, the singing of crickets, the elaborate nest-building by bower birds and the "necking" of giraffes are but a few examples of the amazing variety of courtship behaviours that have evolved.

In spite of the many books written on this subject, it is difficult to generalize about the evolution of these very different reproductive patterns, other than to say that those species that produce a very large number of offspring usually exhibit small body size, rapid development, a short lifespan and live in a rather unstable environment with unpredictable resources. In the case of species producing few offspring that enjoy careful parental care, the opposite conditions often apply, although there are many exceptions. No doubt each pattern has been honed by natural selection to optimize fitness in each species' particular niche.

In any event, we have now learnt enough about the reproductive cycle in red deer to realize that it is a complex process, involving many delicately timed interactions among hormones. In addition, we have seen that the energetic cost of antler growth, fighting among males, courtship rituals, pregnancy and lactation is undoubtedly high. This is not only true for mammals but also for most other animals; consider for just a moment how much energy is required by certain bird species to yolk their eggs, incubate them and then feed the demanding chicks almost to adulthood. Even a tiny mosquito requires a relatively large blood meal to yolk her eggs, and carpenter bees spend large amounts of energy boring tunnels into hardwood to ensure the safe development of their young.

Therefore, if reproduction itself is such a costly process, we could justifiably now ask one of the most difficult questions in biology, "Why add to this cost by evolving sexual reproduction?" Most biologists take the phenomenon of sexual reproduction for granted, or if pressed to answer why it evolved, will usually say something like, "It facilitates evolution and therefore evolution produced sexual reproduction."

This is a highly teleological explanation because it uses the purpose of a process to explain the means by which it arose. Nevertheless, it is difficult to find explanations which are not flawed by this philosophical approach. For example, the standard reason given for the evolution of sex is that the recombination of genes during fertilization results in greater genetic diversity, thereby providing more rapid genetic change and a richer substrate for natural selection to act upon. This would be particularly advantageous in a rapidly changing environment and for developing resistance to parasites, but would confer a greater advantage on the species than on the individual. This is the wrong way of thinking about evolution because natural selection acts primarily on individuals and only secondarily on species.

Arguments of this nature also identify our genes as the major beneficiary of sex; perhaps Richard Dawkins at Oxford is correct in his description of genes as selfish. For example, sex may be just another example of our genes using our bodies in a parasitic fashion to serve their own end or as Dawkins puts it, "DNA neither cares nor knows. DNA just is. And we dance to its music." On the other hand, what about the many animals that reproduce *asexually* very successfully? The small aquatic invertebrates known as bdelloid rotifers, for example, have survived successfully for millions upon millions of years without resorting to sexual reproduction. In fact, asexual females will produce twice as many of their own genes as they would when reproducing sexually. Asexual reproduction even occurs in vertebrates, as in the lizard species *Cnemidophorus uniparens*. These lizards (all females naturally) still engage in courtship and pseudo-copulation to stimulate ovulation – a case of sex without sex!

It is not surprising then that some biologists consider that sex represents a luxurious expenditure of energy. A luxury, it would seem, that most species have learnt to live with, while one particular species has turned it into a recreation, if not an obsession.

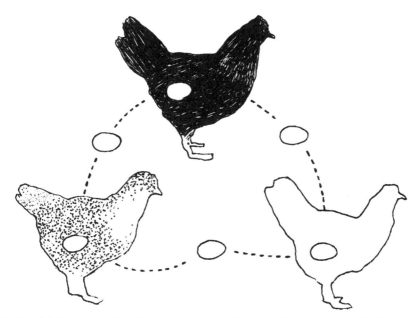

Is the chicken merely a device to ensure the continued survival of the egg?

Finding
the
Way

The ability to find the way can often be of crucial survival value to both humans and animals in their quest to obtain sufficient energy from the environment. Many birds, some large mammals and even a species of butterfly undertake long journeys to exploit seasonal surpluses of nutritional energy and favourable climatic conditions in distant places. Smaller animals have to orient and navigate within their home ranges, and it is essential for social insects to be able to return swiftly to their nests after foraging forays. The latter requires pin-point navigation, if these insects are to make efficient use of their energy reserves. A honeybee that is a poor navigator will use up too much aviation fuel on its foraging expeditions, and become a drain on the colony.

Let me introduce the subject by first looking at how humans cope with navigational problems. In the simplest situation humans will use a mental map of landmarks acquired through personal experience. We all use these mental maps on a daily basis to get to and from work and to perform other important activities. Sometimes we are called on to enter a new, unknown environment and to find our way to a specific place within that environment. Instead of using a cognitive map, we will have to rely on a printed map of landmarks compiled by someone who knows the area in question, or follow a set of instructions based on landmarks described to us by an experienced person.

These situations usually do not present any great difficulties, nor is it particularly difficult to find your way back to home base when exploring unknown territory. Imagine if you were asked to survey the animal life in an unexplored region: all that you would have to do is to continually memorize landmarks and rough distances on the outward journey to fix your position and then to return via these landmarks in reverse order. But, problems do arise when at sea or flying an aircraft in dense cloud, and there are no landmarks available for orientation. In both cases the navigators will make use of a system known as dead reckoning: they estimate their speed from a speed indicator and their direction from a compass bearing and then integrate these two variables to estimate their position. This is the navigational method used by Columbus, when he set sail for the New World in 1492. The problem with dead reckoning, however, is that the ship will experience an unknown degree of drift from ocean currents, and the aircraft from winds, possibly resulting in major errors, often with disastrous consequences.

Today modern satellite navigational aids such as the Global Positioning System (GPS), whereby a ship's navigator can receive a radio signal from a satellite and then read off the ship's position to an accuracy of tens of metres, overcome all these problems. Before the arrival of GPS, naval navigators could adjust for drift as soon as the cloud cover cleared, by taking a fix on either the sun or the stars using a sextant at an accurately known time. These measurements could then be converted to map co-ordinates by referring to a set of navigational tables. The sun will be in a specific position at a specific time, when viewed from a specific place on our planet. If you know the sun's position and the time, it follows that your position can then be calculated. The accuracy of the chronometers would be of critical importance, because a time difference of only a minute represents an error of about 30 kilometres. Consequently, the British Admiralty used to offer large prizes to watchmakers for the production of accurate timepieces for navigation as recently as 200 years ago.

But, how did humans navigate before the invention of the compass and chronometer? Can we use this knowledge to unravel some of the puzzling aspects of bird navigation? Surprisingly little is known about the early navigators, apart from their obvious reliance on landmarks, crude maps and verbal descriptions, as well as the stars and the setting and rising sun. The most successful of the early navigators were probably the Phoenicians. We know very little about the exact methods they used, apart from their reliance on the star tracks. In fact, Greek sailors still call the Pole star the "Phoenician Star." The success of the Phoenicians has been well documented; for instance, their navigation skills allowed them to circumnavigate the entire continent of Africa as early as 600 B.C. The journey took three years and was underwritten by the Egyptian pharaoh Necho.

When the Vikings crossed the North Atlantic to reach North America in 1002, they must have relied heavily on the star tracks at night, which would have been very familiar to them. The rising and setting sun would have assisted them in the early mornings and late afternoons, but what about cloudy days, so common in these latitudes? Danish archaeologist Thorkild Ramskow has suggested that the "sunstones," described in the old Viking sagas, were large crystals through which the Vikings could perceive polarized light. This would have allowed their navigators to determine the direction of polarized light and therefore the direction of the sun, even on cloudy days.

Rüdiger Wehner, Professor of Zoology at Zurich University, reports that a modern aircraft was steered with reasonable precision from Norway to Greenland using a cordierite crystal as the only navigational aid. These crystals are found fairly commonly on the Norwegian shoreline. Several animal species have been shown to have the ability to perceive polarized light and, as we shall see in the case of certain insects, this ability is essential for their navigating success.

We biologists pride ourselves on our navigational abilities. A quick survey of the vegetation should give us a good idea of our general location and altitude above sea level. If we are in a temperate area, we can find out on which side of the rocks the plants grow best. In the northern hemisphere this will be on the south side and vice versa in the southern hemisphere. In a foggy desert the sides of both pebbles and rocks facing the direction of the ocean will show a much more luxuriant growth of lichens. We could also follow streams or even dry river beds downstream in the anticipation of finding nearby habitation. Unfortunately, the truth of the matter is that most biologists today are urbanites and more adept at handling computers than finding their way through the wilderness.

Humans, therefore, are not particularly good or bad navigators. They rely heavily on a good memory and cognitive maps. They are unable to perceive polarized light, their sense of smell and hearing are only moderately well developed and, in spite of recent reports claiming the opposite, they do not appear to be able to use the Earth's magnetic field for direction finding. In spite of these drawbacks, they have colonized almost the entire planet and remarkable feats of "finding the way" are regularly reported. Recently a young yachtsman, while sailing solo from the Cape of Good Hope to Australia, was driven deep into the South Atlantic near Marion Island by fierce storms. All his navigational aids and radio equipment were destroyed. Undaunted, he kept the Southern Cross on his right hand at night and the setting sun at his back, and sailed straight into Perth harbour several weeks later in fine shape.

In contrast to humans, birds are far superior navigators and this is reflected not only in the uncanny feats of homing pigeons finding their way home over unfamiliar territory under a thick cloud cover, but also in the spectacular migratory journeys performed by many species. A single wandering albatross, for instance, has been followed for 33 days by satellite tracking over 15 200 km above the open ocean, before returning to its nest site on a small ocean island. The record for such feats must, however, go to the Arctic tern that covers about 30 000 km per year on its migratory flights. The energetic cost of these long journeys is very high and in most migratory species, the migration is usually preceded by a period of intense feeding and fat deposition to allow enough fuel for the long flight. In the case of tiny hummingbirds that have to fly across the Gulf of Mexico, the fat content of their bodies reaches almost 40 percent. On arrival their fat stores are almost depleted. If unfavourable weather conditions prevail, such as head winds, many birds do not make it and perish in the ocean. Obviously, in most cases, the cost of migration is more than offset by the advantages of enjoying a milder climate, longer days and access to superior nutrition at the destination. But how do the birds find their way? Let us first list the various possible navigational cues that could be used, before analyzing them separately.

The most obvious cue which immediately springs to mind is the sun, and then the stars to provide a basis for celestial (sky) navigation. Polarized light, which will be explained later, is another possibility. The Earth's magnetic field provides a near-ideal grid for obtaining compass headings if the animal in question has the ability to detect geomagnetism. The ability to sense weak odours and low sound frequencies (infrasound) from a distant destination could also assist homing and the fine sensing of barometric pressure would allow birds to select favourable altitudes at which to fly under various weather conditions. We can now briefly review the relative importance of these cues and abilities.

Many experiments have shown that birds are able to orient with reasonable accuracy when they are able to see the sun. If the sun's position is artificially altered by using mirrors, caged migratory birds alter their escape orientation proportionately. Escape orientation is the direction in which the birds try to escape from their cages during a period of migratory restlessness. It is usually recorded by means of an automatic device attached to the cage. These birds employ the sun's direction (the azimuth) as a cue and not its elevation above the horizon.

The sun's azimuth moves across the sky at a rate of change which depends on the time of day, the latitude and the season. Birds must therefore compensate for this change by relying on an internal clock or circadian rhythm, which tells them where the sun should be at any particular time of the day. This dependence on an internal clock has been well demonstrated in homing pigeons by shifting their internal clocks by artificially altering the period or rhythm of daylight to which they were accustomed. After the internal clocks had been altered, and the pigeons had become accustomed to the new daylength regime, the pigeons were released a good distance from their loft; whereupon they disappeared from sight on an incorrect compass heading. The altered angle of their initial heading corresponded reasonably well to the error that could be predicted from the degree of change that had been artificially engineered in their internal clocks.

Birds seem to be able to use the lengths of shadows of various objects to read off the sun's azimuth more accurately than when looking directly into the sun. The after-glow of the setting sun is also frequently employed for orientation by migratory birds and although we still do not know the exact mechanism whereby the sun's direction is used for homing, it is clearly an integral part of many birds' homing systems.

The night sky offers another obvious clue for navigation and the fact that certain species of birds use the stars, or more correctly stellar navigation, has been shown in several ingenious experiments carried out in planetaria. Gustav Kramer has shown that caged migratory birds display a predictable escape orientation beneath the artificial night sky of a planetarium, that corresponds to their normal orientation when migrating naturally. In addition, escape orientation can be artificially altered by changing the star pattern in the planetarium. This is easy to grasp, but how do the birds compensate for the ever-changing star pattern that occurs during the night? Apparently they do not use an internal clock for this purpose, because birds in which the internal clocks have been artificially altered do not make predictable errors in their escape orientation. It seems rather that in some cases they rely on a single star, such as Polaris, which is the star that currently remains closest to the celestial north pole. It can be fairly easily located by reference to nearby constellations. Nevertheless, although stars are important navigational cues in at least several bird species that have been studied, the exact mechanism of their employment is not yet fully understood.

Due to a combination of the movement of the molten iron core within our planet and the earth's rotation, a dipolar magnetic field exists over the surface of the globe. This is a three-dimensional field and provides two kinds of information. Firstly, it possesses polarity and by using a magnetic compass, first discovered by the Chinese in the second century, magnetic north can be found. Secondly, the magnetic field lines leave the surface of the Earth close to the southern

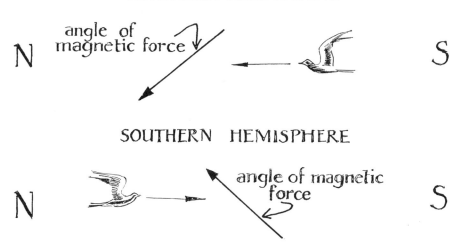

Figure 7.1 – Birds perceive the Earth's magnetic field by the angle which the lines of force make with the horizontal. In the above diagram, both birds are flying towards the acute angle that the magnetic lines of force make with the surface of the Earth; but they are flying in opposite compass directions because of the opposite angles of the force lines in the northern and southern hemispheres.

geographic pole, rise above the earth's surface; and re-enter it close to the northern geographic pole. Consequently, the magnetic field lines form maximum angles of inclination with the horizon at the geomagnetic poles (90°) and a minimum angle at the equator. Therefore, if an animal were to use the polarity of this geomagnetic field to orient, it would have to possess a mechanism which would be similar to a magnetic compass. Alternatively, it could sense the angle of dip of the three dimensional magnetic field rather than its polarity. In this instance, it would distinguish poleward or equatorward directions, rather than between north and south. Research to date suggests that birds use the angle of dip of the magnetic field.

Most of the experiments performed to date on the sensing of magnetism by birds have been of an indirect nature. The escape orientation of caged migratory birds is altered when artificial magnetic fields are applied to the cage. Fitting small bar magnets or tiny Helmholtz coils to pigeons has on occasion disturbed their natural homing ability. Similarly, when pigeons were released near Boston, Massachusetts, in an area where the Earth's magnetic field was irregular, the homing of the birds was jumbled and disoriented until they had left that area. There seems little doubt that at least several species of birds are sensitive to magnetic fields, but the exact way in which the magnetic field is sensed and used is still far from clear. Many of the experiments have not been successfully repeated and the data are frequently highly variable.

In spite of this uncertainty, the search for a ferromagnetic sensing device in birds is an active field of research. The first attempts in this direction were to search for magnetic (Fe_3O_4) particles in the central nervous systems of birds, similar to the particles that have been found in certain aquatic bacteria, such as *Aquaspirillum*. These particles allow the bacteria to "swim" along the inclined lines of the Earth's magnetic field downwards towards the mud, their preferred habitat. This approach in birds has not been very successful and the latest research indicates that many animals capable of sensing magnetism may be using light-sensitive cells (photoreceptors) to sense the magnetic field of the Earth, or more correctly, to sense the degree to which the magnetic field alters the responses of specialized photoreceptors.

The next navigational cue to consider is polarized light; but let me first explain some of the properties of polarized light before evaluating its role in bird navigation. Light consists of electromagnetic energy and it "vibrates" in all directions. When it strikes particles in the Earth's atmosphere, however, it tends to vibrate in a specific direction, thereby forming a characteristic pattern of polarized light in the sky, which changes with the time of day. This can be simply shown by placing a camera, fitted with a 180-degree fish-eye lens and

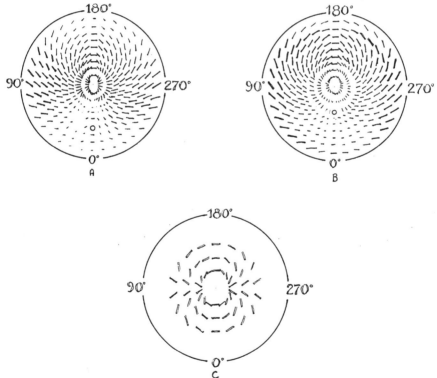

Figure 7.2 – Two-dimensional representation of the polarized light pattern in the sky at a sun elevation of 24° (a) and at 60° (b). The sun appears as a small white disc. Diagram (c) below represents the generalised sky map or template in the retinas of bees and ants, used to analyse the pattern of polarized light in the sky (after Wehner).

polarizing filter, on its back and photographing the sky at various times of the day. From these photographs it will be seen that the pattern of polarization shifts around the sky with the apparent movement of the sun across the sky. The change in the distribution of the angles of polarization, when the sun's elevation is at 24° and 60° above the horizon, is depicted in Figure 7.2. Note that the pattern of polarization changes intrinsically as the elevation of the sun changes during the day.

The plane of polarized light could therefore theoretically be used for direction finding by birds, particularly on cloudy days. Although pigeons can sense polarized light, there is as yet no firm evidence that they use it as a navigational cue. In fact, pigeons fitted with frosted lenses are still capable of flying homeward.

Several other physiological characteristics of pigeons have been invoked to explain their navigating ability. For example, they are able to detect changes in barometric pressure very accurately and able to detect infrasound as low as 0.06 hertz. Humans are unable to detect sound below 10 hertz. This ability may allow migratory birds to orient towards distant sounds such as waves breaking on a shoreline, hundreds of kilometres away. It would, however, not be possible for the birds to pinpoint the source of this sound by triangulation, i.e., by comparing the strength of the separate signals received by their individual ears.

Professor Papi of Pisa University has carried out some intriguing experiments to examine the use of odours or olfactory cues by homing pigeons. Although he has had considerable success in demonstrating that the homing ability of pigeons is disturbed by destroying their sense of smell, most other researchers have not been able to confirm his results. The use of olfactory cues therefore remains a speculative idea and it is difficult to imagine which odours could be used as cues.

To conclude then, the remarkable ability of birds to orient and navigate is a fascinating subject of research which borders on the discovery of the much sought after sixth sense. Nevertheless, we are still far from finding a final answer. The birds seem to be able to use a hierarchy of cues; the sun, stars and geomagnetism at present seem to be the most important and these are employed according to the dictates of weather and other circumstances. The use of landmarks or learned maps is naturally also important; but we must await further research to learn if additional cues, such as odours, infrasound, polarized light and gravity may also be playing an interacting role.

Birds are not the only animals that exhibit accurate homing ability. In fact, compass orientation appears to be a fairly widespread phenomenon in the animal kingdom among both invertebrates and vertebrates. The Saharan desert ant *Cataglyphis* provides an excellent example among the insects. Its navigation ability has been carefully examined by Rüdiger Wehner and this brief account depends

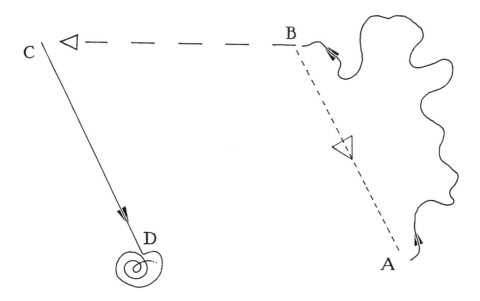

Figure 7.3 – A *Cataglyphis* worker ant leaves the nest entrance A and travels along a meandering foraging path to reach B. The return track from B to A is direct over unknown territory. When artificially displaced from B to C, the ant uses the same compass vector to reach the fictive position of the nest at D, where it commences a circular search.

heavily on his imaginative research. We begin by analyzing the homing ability of these ants as depicted in Figure 7.3.

In Figure 7.3 we can see that the outward bound track of a *Cataglyphis* worker from the nest (A) takes on a meandering pattern as the ant forages for food. In contrast, the homeward track (B⇢A) that passes over unknown territory is remarkably straight. The ants are therefore not following a scent trail back to the nest. Also, if the ant is captured at B and moved to point C, its homeward track would be at the same compass angle (vector) when departing from point B. In the latter case it would naturally miss the nest but, in spite of this, the ant will commence a circular search for the nest after having travelled a distance similar to the length of B⇢A, or the distance of its natural homeward journey, had it not been moved. These results suggest that the ants are using some form of dead reckoning which is cued either by an internal or external referencing system, such as a cue from the sky.

A series of careful and highly ingenious investigations by Rüdiger Wehner and his colleagues has now shown that the most important cue used by these desert ants is polarized light. Apparently the ants possess a programmed map or template in the retinas of their eyes. The anatomical arrangement of the light-sensitive cells (photoreceptors) in the retina provides the basis of this map, because these photoreceptors are arranged in such away that they match the pattern of polarized light in the sky (see Figure 7.2). The retinal map can therefore serve the function of a so-called "matched filter." For example, if the ant holds its head at a constant angle to the horizon, it can scan the sky by sweeping its array of photoreceptors in the retina across the polarization pattern of the sky. It does this by interrupting its homeward journey with occasional, but rapid, 360° turns, described by Wehner as "graceful little minuets." When the retinal map matches the map of polarized light in the sky, maximum brightness will be received by the retina.

This, then, is how the ants derive their compass information from the pattern of polarized light in the sky, but because the pattern of polarization rotates with the apparent westward movement of the sun across the sky, the ants must compensate for this change by reference to their own circadian or internal clock.

We can now understand how a foraging ant could continuously update its homeward compass heading, by scanning the sky regularly, but how do they measure the distance they have travelled so as to know when to stop and start a circular search for the nest opening? At first there was no clear answer to this question. The first suggestion to be made was that the ants were somehow measuring their energy expenditure on the outward journey, but this cannot be true because it has been shown that they are able to measure the distance they have walked irrespective of the size of the load they are carrying. Wehner points out that *Cataglyphis* ants maintain a constant rate of movement while foraging and that they may somehow be able to store information about the number of steps they have taken on their outward journeys. A more favoured explanation is that they are able to visually record the 'movement' of the sand surface (and therefore the distance travelled) as they walk over it at a constant height and almost constant speed. Perhaps "sprint" is a better term than "walk" as these little chaps can cover a distance equal to 100 of their body lengths per second – even a cheetah cannot match this performance.

To conclude then, although many animals are able to orient and navigate well, we still only have partial answers to the question of how they achieve these remarkable feats. Many fine minds are at present focusing on these questions, and we can expect even more fascinating results as this scientific detective story unfolds.

African Safari and the Sprinting Compost Heap

We have examined the energetic costs of various kinds of locomotion; of keeping warm; of reproduction and the foods that are available to fuel these energy requirements. We can now apply this knowledge by exploring a natural ecosystem to find out how energy flows through such a system. At the same time we shall see what an extraordinarily important role cellulose plays in providing energy within a natural ecosystem.

The African savannah is one of the most exciting ecosystems on our planet to visit and study and, fortunately, there are still some large tracts of unspoilt savannah left in southern Africa to enjoy; so let's meander through an unspoilt area of bushveld, as the savannah is called here, and see what we can learn.

Our first impression, at any season of the year, is one of astonishment at the large variety of plants: mostly grassland, but interspersed with trees, shrubs and herbs, and a riverine forest along the banks of a dry river bed. All contain large amounts of cellulose. Equally impressive is the rich variety of animal species living within this plant community, in spite of severe seasonal droughts.

The first animals to catch our attention will probably be the birds, which exist in profusion. Fish eagles cry from their high perches above water, crimson breasted shrikes flash in the woody thickets and large ground hornbills strut about as they make their hooting sound. Small flocks of minute white-eyes are gleaning insects from branches and leaves as they flit through tree canopies and, overall, there is the rich drone of abundant insect life.

We may be fortunate on our first afternoon to experience an impressive thunderstorm. The crisp white cumulus clouds have been building all afternoon into thunderheads, the landscape is still brightly lit by the quartering sun and a small herd of impala antelope passing by is highlighted against the deep purple of the approaching storm. Soon the sky darkens completely, the wind picks up, and the thunder and lightning reach a crescendo as a brief deluge swamps the veld. Dry washes run with muddy water for a short time; the sky clears rapidly and the rain water on dry, hot soil produces a malty aroma, adding to the enjoyment of the freshness of the air after the rainstorm. Pliny, the famous but over-imaginative Roman naturalist, described this aroma as the incomparable sweetness which the earth conceived from the sun.

The rain has also caused a mass exodus of winged termites from the many termite mounds dispersed across the veld and, in response, a large flock of greater striped swallows is sweeping low over the vegetation as they hawk for insects. The birds often engage in steep, almost vertical climbs to capture their prey and these manoeuvres are followed by stall turns with closed wings, that allow them to enter a shallow dive to pick up speed again. Their swift flapping flight is frequently interrupted by fairly long glides and beautifully controlled steep turns with wings at 90° to the horizon – a manoeuvre that would cause a fighter pilot to black out without a G-suit. They slowly disappear, leaving a memory of delta wings and swallow tails that delight the eye with their functional efficiency.

The next birds to attract our attention just after dusk are double-banded sandgrouse, flying in ragged little squadrons emitting their characteristic call of *oh NO, he's gone and done it aGAIN* repeatedly. They wheel carefully before landing at the water hole, where they remain constantly alert to escape predation.

We wait patiently at the water hole for the elephants, and our first visitors are a small pride of lions that drink long and deeply from the muddy water. Lions can, if necessary, do without drinking water for several days at a time, and those living in the Kalahari Desert for even longer. This is largely due to their efficient kidney function which allows them to produce a highly concentrated urine, thereby reducing water loss significantly. Our interest in the lions is abruptly ended with the sudden but silent arrival of a herd of elephants out of the darkness. Their huge padded feet make almost no sound on the hard-packed earth of the game trail they have used to approach the water. We suddenly become aware of a huge shadowy presence, before making out the outlines of several large females as they move around the water hole, shooing the lions away. The lions slink off reluctantly and the elephants begin their bathing and drinking rituals with obvious pleasure and much social interaction within the herd.

Once the elaborate bathing rituals are completed, the adults congregate and stand very quietly side by side for a long time as they rest after a long hot day of foraging. From time to time you will hear loud rumbling sounds emanating from the animals and if you interpret these as gigantic stomach rumbles, you will be correct. What you will not hear, however, is the low frequency infra-sound, on the edge of the range of human hearing, that elephants also produce. These sounds, similar in frequency to those made by some whales, travel long distances and are probably used for communication between sexes and different family groups. Infrasound produced by elephants can even penetrate dense bush up to distances of five kilometres.

The gigantic stomach rumbles we heard are caused by large amounts of gas produced by cellulose digestion in the digestive tracts of the elephants. These animals consume the most fibrous diet, with perhaps the exception of the termites, of all the bushveld animals. A zoo elephant can delicately pick up a peanut from your hand and convey it to the massive crushing molars in its mouth, but in the wild, elephants depend mostly on coarse plant material and they can demolish large trees, shrubs and grassland if necessary to obtain their nutrients and energy. This highly fibrous material is only partially masticated before passing down the gullet to the true stomach. Here the highly digestible portion of the food is prepared for digestion and absorption in the small intestine. But only a small percentage of the food is digestible; the bulk of the ingesta moves to the greatly enlarged caecum, part of the large intestine, where its passage is slowed and the fibre-rich food undergoes digestion by millions of micro-organisms, mostly bacteria.

The biochemical changes which occur in the caecum are exceedingly complex and involve the production of simple fatty acids, the synthesis of important B-complex vitamins and some important amino-acids, which are the building blocks of proteins. Many of these compounds are of critical importance in the diet of the elephant and they are absorbed through the caecal wall. The fatty acids will provide energy, much like sugars do; and the B-vitamins will take part in important enzymatic reactions. Truly a remarkable process for turning crude, rough plant material into refined and valuable nutrients. In contrast, as we have seen, a tiny mosquito feeding on the blood of an elephant has all the nutrients prepared in a highly digestible form.

Caecal digestion is by no means a complete process and this is evident in the large elephant droppings, which still contain a great deal of fibre, but also enough nutrition to attract swarms of dung beetles. Africa is fortunate in having many species of dung beetle to process and remove the dung of hoofed animals to underground nests, before flies begin breeding in large numbers in the dung. Australia is not as fortunate because the absence of hooved animals on that continent for millions of years has meant that indigenous dung beetle species have not evolved the ability to process the dung of hooved animals. The large scale introduction of cattle and sheep into Australia has consequently resulted in widespread plagues of flies, and any outdoor activity in the interior of Australia becomes far more pleasant after sunset.

Next morning we decide to explore our environment more systematically by tracing how energy flows through some of the major food chains in this region. We should begin with the surrounding soil types and study the complex ecology of root systems, which is of great importance but is a neglected field because of the practical difficulties involved. Our time, however, is limited and we can reflect only briefly on the complexity of soil ecology in all terrestrial ecosystems. Soil fertility is maintained by countless interactions among the myriad of soil organisms, the surface vegetation, the herbivores grazing the vegetation and the climate. Clive Ponting informs us in the *Green History of the World* that just one acre of good soil in a temperate region could contain as many as 125 million small invertebrates and that 30 grams of that soil will contain one million bacteria of just one type, 100 000 yeast cells and about 50 000 fungal mycelia. Staggering figures but easily confirmed by a few weeks study with a microscope and some agar plates. Should you undertake such an investigation to the west of our present study area, you would find a specific fungus growing in association with one of the bushveld grasses that tastes similar to the prized truffles of France. In fact, a small industry to export these fungi to France has just begun to develop in the semi-arid savannahs of Namibia.

We move now into an area of parkland dominated by a variety of grass species with imposing Latin names like *Eragrostis*, *Panicum*, *Digiteria* and small patches of the highly palatable short grass *Cynodon*. These grasses, together with the herbs, shrubs and trees, represent the primary producers of our ecosystem, as they are responsible for capturing the electromagnetic energy from the sun to produce sugars rich in energy. It would take several lifetimes of many ecologists to unravel all the interactions among the many species of animals and their food plants in the bush-veld. So let us examine just one species of grass, buffalo grass, to confirm this supposition; this will lead us to the second link in the food chain, namely to the herbivores.

Our grass species grows almost waist-high and, in addition to a soft stem, possesses several palatable leaves. It is one of the favourite food plants of the graceful impala. Impala are true ruminant animals. They graze upon a variety of bush-veld grasses and browse upon the leaves and fine twigs of shrubs and trees. They have no upper incisors, only a tough horny plate known as the dental pad. When they graze they curl their tongues around a grass tuft, clasp the tuft between the lower incisors and the dental pad, before snapping their heads back with a jerking motion to cut off the upper part of the grass tuft. In contrast, the small herd of Burchell's zebra nearby are almost exclusively grazers, and prefer feeding on short grasses. They have strong mobile upper lips to manoeuvre the short herbage between their upper and lower incisors, where it is cut free. In this way competition between zebra and impala, even when feeding on the same grass species, is reduced.

There are other important differences in the nutrition of impala and zebras. The zebras, like elephants, are not ruminants and therefore do not chew the cud. Instead, like elephants, rhinos and other non-ruminant herbivores that live off large amounts of cellulose or crude fibre, a zebra chews its food well before swallowing and relies on an en-larged caecum for final digestion; this functions in a similar manner to the elephant's caecum. Impala, on the other hand, swallow their food with a minimum of chewing and then, once they are replete, they rest and return the food to their mouths for rechewing mouthful by mouthful.

The impala's large forestomach, or rumen, contains a mass of fermenting plant material and the tens of millions of micro-organisms attacking this moist mass are enjoying an almost perfect physical environment. For example, the acidity, or pH, of the rumen contents is kept constant by a copious secretion of litres of saliva, containing sodium bicarbonate. It is also kept at a constant body temperature of around 39°C, while being continuously mixed by strong contractions of the muscular walls of the rumen; and, although the rumen itself secretes no enzymes, the enzymes of the micro-organisms soon begin to work their miraculous chemistry. Similarly to the zebra and elephant caecum, fibrous cellulose is broken down to simple fatty acids, including acetic acid, the main component of household vinegar. The B-complex vitamins and microbial protein are also synthesized for the micro-organisms' own use, but, as this fermenting mass passes through the true stomach and small intestine, rapid digestion and absorption of these valuable nutrients takes place. Again, the changes in the rumen, which are not dissimilar to the changes which take place in an ordinary compost heap, have allowed masses of crude plant material to be transformed into valuable, refined nutrients; in the case of a nursing antelope, even to milk, one of the most nutritious of foods.

Ruminant digestion is commonplace on the African savannah as it is employed by the many spectacular species of antelope such as steenbok, impala, bushbuck, waterbuck, kudu, and many other species living in this region. Nevertheless, an interesting question is whether caecal digestion is superior to ruminant digestion, or is the opposite true? Ruminant digestion has the advantage that the ingesta is first fermented in the rumen before passing through the more conventional process of digestion in the true stomach and small intestine. In this way, absorption of the end-products of rumen digestion is more complete than is the case with caecal digestion, which occurs near the terminal portion of the digestive tract. Some small mammals that use caecal digestion, such as hares, compensate for this disadvantage by recycling their faeces. At night a special soft faecal pellet is produced for ingestion.

The apparent superiority of ruminant digestion over caecal may also explain why there are so many ruminant species among large herbivores, compared to caecal digesters. For example, only a handful of equine (horse-like) species have evolved, compared to the very large numbers of ruminants distributed throughout the world. This subject has been thoroughly explored by professor Reino Hofmann (1989).

We return now to our original buffalo grass and find that it is not only impala and zebras that are feeding on it. Grasshoppers are feeding on the more succulent leaves and a darkling or tenebrionid beetle is pulling out old stems from the tuft; the bite marks on other stems, together with the presence of characteristic faecal pellets around many of the tufts, suggest strongly that scrub hares are also feeding on a particular part of our grass. Large shiny, black harvester ants are carrying the grass seeds away to their nests to provide a storehouse of energy; and a group of termites is cutting up and removing dried out grass stems to their cool and moist underground caverns.

The small size of the termites is misleading; several ecologists believe that the largest amount of cellulose or fibre in a bushveld ecosystem is channelled through the digestive systems of millions upon millions of termites. The termites, like almost all animals, do not possess their own enzymes to digest cellulose. Instead, as in ruminants, they rely on the enzymes of micro-organisms inhabiting their digestive tracts to digest the vast quantities of cellulose passing through their colonies. During the process of digestion large amounts of methane gas are produced and some ecologists have half-jokingly calculated that this amount of methane, because it has a strong greenhouse effect, could affect the earth's climate significantly. So much so, that sea levels could rise and flood coastal cities!

After this rather superficial study of buffalo grass, we can conclude that it is not used only by one or two species, but provides several different nutritional niches to a number of very different animal species. Ecologists have borrowed the term niche from architecture, but in ecology it has a functional, rather than a spatial meaning, because it describes the *profession* of an organism within the ecosystem. In other words – what it does for a living. Whereas the *address* of the organism is usually described by the term habitat. If we looked even closer with a magnifying glass at our grass tuft, we might find even more animal species, such as nematodes in the roots, or tiny parasitic wasps using the small nodules or galls in the grass stem as temporary housing. All these observations seem to suggest that direct competition for energy among animals is not as common as we might expect and that evolution has produced a surprising degree of separation among the niches occupied by animals. We have had a long hot day and so we gladly retire to our campsite for our evening meal. On this occasion, we may even bake fresh bread, using a hollowed-out termite nest covered with glowing coals as an oven.

After dinner we hear the carnivores, eager for the hunt; the coughing, grunting and roaring of the lions, followed by the maniacal laughter of the hyenas in the distance. Occasionally jackals make their presence felt with high pitched yodelling and whining. Our flashlights pick out a handsome young male nearby that has been attracted by the aroma of our grilling venison. Talk turns to the big cats and the carnivores in general. Someone asks if there are more plant species than animal species on our planet and then the question arises why there are so few carnivore species compared to herbivore species. The first question is not easy to answer. It has been estimated that there are about 250 000 flowering plant species on Mother Earth, but probably several million animal species, with many more still to be discovered. We have also seen how this could be possible from our observations on the multiple uses to which our grass species is put. The reason why there are far fewer carnivores than herbivores is easier to answer because, as we move up a food chain, only about 10 percent of the energy is transferred from one link to the next. Ninety percent is lost at each link in maintaining the metabolic rate of the animals and is largely lost as heat. This means that the original energy produced by the plants becomes so attenuated at the top of the food chain that there is just not enough energy to support large numbers of carnivores, and even less for the few top carnivores. It is also the main reason why your tenderloin steak costs more than ten times as much as the cabbage at your supermarket.

The next day we move away from the parkland savannah we have been studying towards a more wooded area and, eventually, to a riverine community on the banks of a dry river bed. Moving through the woodland savannah we are again struck by the rich variety of bird life obtaining their energy and nutrients from insects, seeds, fruit, small rodents and reptiles. We have to move very cautiously so as not to disturb a large female black rhino with her young calf. She is feeding on the very thorny branches of an *Acacia* bush. The way in which she avoids impaling her tongue on the long thorns is amazing. The fine, green *Acacia* leaves she strips from these branches are an excellent source of digestible energy, providing proteins, vitamins and sugars for the nursing mother.

Our path takes us past a small herd of giraffes walking slowly through the woodland and feeding off the topmost branches of several *Acacia* species. Like the rhino they are adept at stripping the delicate green leaves off the branches, without suffering injury from the formidable thorns. They also enjoy the advantage of being the only hooved animals tall enough to reach these succulent rewards at the top of the trees. The giraffes soon become aware of our presence and they move away with a slow, but rather dignified, gait.

We are now approaching the riverbank and walking through fairly dense riverine forest. The leaf litter stirs with some interesting beetles, but we are intent on finding one of the smaller antelope species in this area, namely the steenbok, that weighs only about 10 kilograms. Steenbok are mixed feeders, preferring fresh green leaf material from the surrounding trees, forbs and shrubs. These leaves often contain a considerable amount of moisture and consequently these small antelopes do not have to drink water regularly, an exercise which can expose them to predation. They also avoid the strong sunlight of the open savannah, preferring the shade of the woody thickets. Their small size also means that their rumen capacity is not sufficient to process large amounts of bulky, coarse material. Instead they rely on a much faster rate of passage through the digestive tract and, because of this, select material with a much lower cellulose or fibre content. This is fortunate because small animals suffer from the additional disadvantage of having a significantly higher resting metabolic rate per unit body weight than large animals.

Our disappointment at not sighting a steenbok is soon forgotten as we come upon a magnificent wild fig tree on the bank of the dry riverbed. The tree is laden with fruit and there is a large troop of vervet monkeys exploiting this resource, rich in energy and vitamin C. The troop chutters and vocalizes incessantly, as information is passed haphazardly among its members. This method of information exchange appears to be far more random than the more structured exchanges of a baboon troop, which we encountered on a previous occasion. Our attention is soon attracted to two black eagles circling close above the top canopy of the fig tree. The monkeys have also seen them and crash pell-mell into the adjoining trees, scolding and squealing their displeasure. One of the eagles swoops on a straggler and while the monkey's attention is riveted on this eagle, the second bird dives from behind to snatch the unfortunate primate into the air with its powerful talons. Were the eagles co-operating in a dual strategy or tandem strike to create a diversion? Perhaps, but more research is needed to confirm this.

All is now quiet as we lunch in the shade of the massive fig tree, except for the drone of African killer bees as they work around the sugary exudate on the figs. The honey of bushveld bees is often very dark, almost the colour of molasses, but exquisitely flavoured. Our conversation now turns to what we have learnt from our short visit to the bushveld. Our main impression is one of the unexpected variety of species and the complexity of the interactions among the plants and animals. But, in spite of this complexity, we have also been struck by the large number and variety of niches available to plants and particularly to animals. How could this intricate web of energy transfer between plants and animals have developed? We all agree, although some rather reluctantly, that it is far too complex to have been designed and could only have evolved over millions of years through trial and error, or to use a more acceptable evolutionary phrase, by chance, opportunity and necessity.

Another lasting impression we have of the bushveld is the key importance of cellulose or fibre digestion for providing energy within the ecosystem. The plants produce tonnes of cellulose per hectare but virtually none of the animals possesses the enzyme cellulase to break down this abundant form of energy to simple sugars. Instead, many have evolved capacious and complex digestive systems to house what almost amounts to a compost heap, in which masses of teeming micro-organisms do the work for them. Why this evolutionary pathway should have been followed, rather than the acquisition of their own cellulase enzymes, is a mystery. Perhaps the fringe benefits of rumen digestion, namely the synthesis of essential amino-acids and B-vitamins, have outweighed the advantage of possessing one's own cellulase. Also, a ruminant enjoys the benefit of being able to graze rapidly and swallow its food almost whole. It can then retreat to a favourable micro-climate where it can rechew its food in peace, while keeping a wary eye open for predators. Whatever the answer to this paradox is, you will now know when you next see a ruminant such as an African gazelle or American pronghorn antelope, sprinting at full speed, that it is being powered by a rather humble compost heap.

Desert Sands

Next we visit a desert biome on the southwestern coast of Africa in the company of several scientists to apply our knowledge in an unusual environment. The group includes a physicist, an ecologist, a botanist, two zoologists and a physiologist.

We look down from the Great Western Escarpment onto the endless desert plain stretching westwards to the cold Atlantic Ocean and southwards towards the largest dune field on our planet. The desert plain is interrupted by *Inselberge* or island mountains and from this height we can clearly see the washes or wadis that have scarred the plain to produce a remarkable pattern of erosion, etched by the sparse vegetation growing along their margins. The colours of the landscape are mostly pale pastels, beige and creams merging into the blue and light mauves of the distant mountains and the empty blue of the desert sky.

We decide to continue our journey down the escarpment to find a suitable camping spot and it is not long before we find one near a water hole. We avoid camping directly beneath the *Acacia* trees surrounding the water hole, in order to keep clear of the bloodsucking tampans, tick-like creatures. These animals live just beneath the sand surrounding isolated trees on the margins of the desert. When large mammals, such as oryx, mountain zebra or humans, seek shade and rest beneath the trees, the tampans sense that the carbon dioxide content of the air has increased slightly, due to the exhalations of the mammals. They then emerge from beneath the sand and attach themselves to the host animal. The tampans possess a highly allergenic saliva and after one or two bites, humans frequently become hyper-allergic to their bites, which can have fatal consequences. But far more remarkable is the extreme sensitivity of tampans to slight increases in the carbon dioxide concentration of the atmospheric air surrounding the tree. This sensitivity has been exploited by biologists when collecting them. They merely place a block of dry ice (solid carbon dioxide) on the sand surface and tampans march in from all directions to investigate this super stimulus and meet a frozen death at -80°C in the desert.

One of the large *Acacia* trees near our camp is weighed down by a huge communal birds' nest of the sociable weaver, *Philetairus socius*. At first it seems that the nest, measuring 6 x 4 metres, is a single large nest, but closer examination reveals that there are about 40 individual nest chambers. Although the birds assist one another in building the communal nest, they are not truly social, like the ants and bees, because they breed independently. Nevertheless, they all enjoy the excellent thermal insulation which the nest provides; it has been shown that the nest temperatures are far less extreme than the widely fluctuating outside air temperatures typical of the desert. The nest is mostly constructed of dry grass stems with their sharp points pointing downwards into the individual entrance tunnels on the underside of the nest. This configuration makes it difficult for predators to enter a nest. Nevertheless, the Cape cobra succeeds frequently in removing eggs and young from the nests.

We spend the last few hours of light patiently watching various activities at the nearby water hole and eventually we are rewarded by a lanner falcon striking down a sandgrouse in midair in an explosion of feathers. With typical esotericism, our physiologist comments on the explosion of adrenaline in both the falcon and the grouse, describing it as a catecholamine storm!

Night envelopes us rapidly on this African plain and the barking geckos emerge from their shallow tunnels to begin their clicking chorus. Our physicist enthralls us with descriptions of the brilliant night sky and, later, someone else plays classical guitar music to complete the magic of our first night in the wilderness.

Next day breaks cool and windless and we begin to explore the wash next to our camp moving as quietly as possible. We emerge from a sharply eroded bend in the wash to find a group of ostriches grazing the pale bleached grass that has not received any rain for over two years. One wonders what benefit they could be obtaining from the lifeless grass. However, Mary Seely, Director of the Namib Desert Ecology Research Unit, and I have shown that these grasses are hygroscopic. In other words, during the early hours of the morning when the air is cool and the relative humidity rises, they can absorb water vapour from air that is not even saturated with water vapour. Just before sunrise this seemingly dry grass can contain as much as 24 percent moisture. Therefore, by limiting most of their grazing to the early hours, the ostriches are ensuring that they take in appreciable amounts of moisture with their food, which is of great benefit in a desert environment. Although the dry brittle grass appears to have very limited nutritional value, in fact it is much like a naturally cured hay and is still an important source of energy for the birds. Also, ostriches are probably the best digesters of crude fibre or cellulose among all bird species.

On our return journey to the camp we can feel the desert heating rapidly and soon the lappetfaced vultures will be sailing on the thermals above the plains. Their keen eyesight will scan the desert carefully for carrion as they make endless gentle turns, controlling their flight effortlessly with small adjustments of their primaries, tail feathers and wings. We startle a black-backed jackal, probably returning from a nearby *Inselberg* where it has been hunting nocturnal desert rodents. At first it is oblivious to our presence, as it lopes along with a springy gait to minimize energy expenditure. But soon it stops, rotating its head to analyze the scents drifting slowly towards it on this windless morning. It satisfies itself that all is well and continues towards its den, where it will hole up during the heat of the day. On its way home the jackal will keep a keen lookout for the eggs of ground-nesting birds such as the sandgrouse or the odd lizard and, if very hungry, will not overlook tenebrionid beetles.

Our physiologist reminds us that jackals belong to the dog family and consequently enjoy the benefits of a remarkably sensitive sense of smell. He then treats us to a brief talk on the sense of smell by reminding us that trained police dogs can follow the tracks of their handlers, even after dozens of strangers have stepped in exactly the same tracks as the handler. They can even pick up the female sex hormone progesterone on objects handled once by women in the second half of their menstrual cycle.

Why then, we ask ourselves, have humans a poor sense of smell, comparatively speaking? Perhaps, if we had evolved a very sensitive sense of smell, the world would become too busy an environment for us. Too much information would reach our brains which, in pre-historical times, would perhaps have been more gainfully employed by grappling with rational problems like how to build a shelter, or grow food, or outwit predators. Presumably, suitable neural filters could have evolved to protect the brain from an information overload. In any event, we have survived and reproduced successfully for millions of years, apparently without a super sense of smell. It would seem then that, in the absence of necessity, selection pressure for this attribute has not been strong.

This does not mean that our sense of smell is trivial. Perfumers, wine connoisseurs and tea tasters have a remarkable ability to distinguish among aromas and, more subtly, we may subconsciously be receiving and reacting to many olfactory stimuli. These may govern our interactions with people in important ways, such as instant dislikes and attractions.

Before reaching camp one of the party finds a stone cricket among the quartz pebbles strewn across the plain. The camouflage of the cricket mimics the quartz stones almost perfectly. Excellent camouflage or crypsis in desert animals is critically important because of the scarcity of vegetation in which to hide. The quartz stones themselves are also of interest, because one finds algae growing on their lower surface. The algae obtain moisture from regular fog off the nearby Atlantic Ocean, which occasionally condenses on the surface of the stone and runs down to the lower surface. The quartz pebbles are also sufficiently translucent to allow enough light to pass through them to support photosynthesis in the algae. Nearby opaque pebbles have no algae beneath them. We then examine the algae beneath a quartz pebble with a magnifier and find tiny lepismids or silver fish living in this moist and cool microhabitat – a veritable rain forest in the middle of the desert for these tiny creatures.

The special clarity of early-morning light in the desert has passed now and it is becoming increasingly hot. The gravel plains are beginning to shimmer through our binoculars and, off to our right, we notice that a small group of ground squirrels, *Xerus inauris*, have flared their long bushy tails and folded them over their backs to act as parasols against the fierce sun. They also turn their backs to the intense radiation to enhance the shading effect while they forage. We clamber into our Landrover and lurch and sway along a rough track towards the eastern periphery of the desert, in search of an area that has reportedly received some rain. Precipitation in these areas falls most often in isolated patches in the form of brief thunderstorms, so that it will be difficult to find.

On our way we make contact again with the group of ostriches we saw earlier. On the edge of the group is an adult female that is nervously pacing back and forth. We soon see why; she is in charge of a large crèche of chicks, milling around her legs. Suddenly she sprints away, running in a wide semi-circle away from the chicks; about every 30 metres she leans over and drags her wing along the ground. This is the famous broken wing dance of the ostrich, designed to lure predators away from the chicks. Meanwhile the chicks have responded to her behaviour by "playing possum"; they have adopted a frozen posture, lying close to the ground and, in this position, are very difficult to find because of their superb camouflage at this age.

The remaining adults seem unperturbed and we watch how they manipulate their wings and feathers to reduce heat gain from the now intense solar radiation. The feathers on their backs have been erected to form a thermal shield against the incoming radiation. They have turned to face into the sun to reduce the profile area exposed to the direct rays of the sun; and have moved their wings forward and away from their naked chests. The naked skin over the chest area is now in the shade and acts as a thermal window for losing heat by radiation and convection. The ostriches have not started to pant yet, but when air temperatures reach a certain threshold, their respiration rate will suddenly rise from four respirations per minute to forty-four per minute to increase evaporative cooling by thermal panting.

We get the impression that the ostriches' temperature-regulating behaviour is designed to delay the onset of panting for as long as possible, thereby saving precious water reserves within their bodies. Physiologists have also shown that when ostriches are not panting, the air they expire is not fully saturated with water vapour. This represents a further important water-saving mechanism that is absent in humans. We expire air fully saturated with water vapour. You will have noticed on cold days how this water vapour condenses immediately, giving the impression that we are exhaling clouds of steam.

The survival of ostriches in the desert, frequently for long spells without drinking water, is a good example of an important principle in physiological ecology; namely that survival in an extreme environment is often due to a series of rather minor adaptations acting in concert, and not to a single dramatic adaptation.

The chicks we have just seen are proof that the ostriches, even though they are ground nesting birds, can breed in mid-summer. The huge eggs are laid in a mere scrape on the surface of the desert. They are incubated during the night to keep them warm, but during the day they are also incubated to keep them cool or, more correctly, to prevent them from overheating. In the latter case the ostrich covers the eggs completely and engages in rapid thermal panting to prevent its own body from overheating, thereby protecting the eggs simultaneously. If a cool breeze from the cold Atlantic should spring up during late afternoon, the incubating ostrich will turn in the direction of the wind and rise up on its hocks to allow the wind to cool the eggs convectively.

The sandgrouse also lays its eggs in a shallow scrape on the surface of the desert plain, but can only breed in the cooler winter months because of its much smaller size and lower thermal inertia, compared to the ostrich. An adult ostrich weighs about 90 kilograms, whereas sandgrouse weigh only around 180 grams, hence the large difference in inertia. Even more spectacular is the ability of the minute Gray's lark (20 g) to nest successfully on the ground. In its case, because of its small size, it is obliged to nest in the shade of a sparse grass tuft or overhanging rock. The nest cup is also insulated with fine dry grass, which retards heat flow from the surrounding soil into the nest.

Our track swings back to the eastern margin of the desert and someone points out how the huge cumulus clouds are building over the escarpment. A sudden downpour over the escarpment can result in flash floods in the desert and we shall have to take care when we reach the major Kuiseb river system. After an hour of travel, we find the area that received some rain about 10 days ago.

The transformation is dramatic; the perennial grass tufts have already sprouted strong green shoots, many different annuals have germinated, shrubs have suddenly come to life and, as far as the eye can see, fresh green growth predominates. Many animals have started gathering in this temporary Garden of Eden. There are oryx on the far horizon, springbok close by and in the middle distance small family groups of mountain zebra. Zebras, like their cousins the horses, sweat freely in response to high temperatures and hard exercise. Not surprisingly, they are dependent on daily access to drinking water, unlike the oryx that can, if necessary, roam indefinitely without having to drink water. The zebras will therefore return to a distant water hole tonight and waste considerable energy in doing so. The springbok, like the oryx, can survive for long periods without drinking and the fresh green grass it is presently grazing will be more than sufficient to meet its water requirements.

Someone in our group comments on the pale ventral surfaces (or bellies) of the large mammals. In the case of the springbok it is a brilliant white. Several other species of African antelope have white bellies but most deer in the cool climates of the northern hemisphere do not. We conclude that the white colour has evolved to assist in the reflection of solar radiation bouncing off the surface of the soil, thereby reducing the heat load on these animals.

Our track continues along the foothills of the escarpment and we intercept a large flock of sheep being herded towards the greening pasture. These sheep are well adapted to the desert; their thick fleece acts as a thermal shield, keeping heat out. The tips of the fleece reach temperatures as high as 80°C but the surface of the skin remains at a relatively cool 39°C. The sheep have large fat tails. The energy stored in these tails can be mobilized during periods of poor nutrition, a frequent occurrence in desert areas. The deposition of the fat ectopically in the tail and not uniformly under the skin, as is the case in temperate zone sheep breeds, also facilitates heat-loss from these animals.

Winged termites have left their nests in the thousands in response to the rain. Many have died and the local lizards are tanking up on this luxury; bat-eared foxes will soon follow suit when evening approaches. We have noticed that several temporary ponds have formed in depressions on the surface of the soil, as a result of the rain. Within a matter of hours after rain, algae, bacteria and protozoans appear in these ponds and after 2-3 days a variety of small crustaceans appear and start to reproduce rapidly. The pond is literally teeming with life. There are fairy shrimps, clam shrimps and tadpole shrimps, all driven by their DNA to reproduce as frantically and swiftly as possible. Soon the pond will dry out and the eggs, left behind by many of these creatures, will have to tolerate life in a desiccated state for many years until the next rainfall event. For example, the eggs of the tadpole shrimp, *Triops granarius*, can survive 16 hours of exposure to 98°C; and at normal desert surface temperatures between 0 and 60°C, they will survive almost indefinitely.

It is past noon and the day is turning into a real scorcher. Most animals are retiring to various kinds of refuges to escape the mid-day heat, so we decide to drive south towards the dry river bed of the Kuiseb River, which marks the northern edge of the massive dune field of the southern Namib. On the way we stop in the shade of some Euclea trees, growing on the edge of a wash, for lunch. Before long we notice the strong, putrescent smell of rotting meat and investigate among the shrubs and trees to find its source. It originates from a tree in full flower; the flowers are producing the strong scent to attract flies, which then serve as pollinators. The tree is aptly named *Boscia foetida* and closer examination reveals a large desert chameleon, *Chamaeleo namaquensis*, catching many flies attracted by the rotting odour. Bryan Burrage studied the behaviour of these large desert chameleons and found that their body temperatures could range from 1°C in the early morning, after a particularly cold night, to as much as 40°C just after mid-day, when the cold air had been displaced by a strong east wind off the escarpment. These chameleons possess nasal salt glands which produce a highly concentrated salt solution to rid their tissues of excess salt with a minimum loss of water, thereby assisting their rather inefficient kidneys. Hence the white accretions of salt frequently seen around their external nostrils.

On the surface of the wash we notice a column of rather large, shiny black harvester ants ferrying small seeds back to their nest. We comment on the large number of social insects on this desert plain and how well suited they are to desert life. Their flexible life style includes the efficient division of labour, that allows them to build well-protected, cool, humid nests below the ground, in which they can store food and escape from the hot desiccating environment on the surface. When times are hard the colonies can shrink to a skeleton size, so to speak, and then expand rapidly when favourable environmental conditions return. For most of the colony's life, the members are all female, even the soldiers; males are produced only when required. The capabilities of ants and termites are truly remarkable and include the practice of agriculture, architecture, cannibalism, slavery, chemical warfare and even navigation. No wonder then that, after hearing about all these attributes during a lecture given by the distinguished myrmecologist Rüdiger Wehner, some-one in the audience asked, "If ants are so smart, why have they not discovered fire and the printing press?" Rüdiger replied (tongue in cheek), "There is a minimum size for the existence of a flame and that is too large for ant use; moreover, in books of a suitable size for ants, the pages would stick together."

Another intriguing example of how well the social lifestyle is adapted to life in the desert is provided by the naked mole rat. Jennifer Jarvis of Cape Town University was the first to discover that these ugly little creatures are truly social, or more correctly, eusocial mammals.

To earn the description of being eusocial, a species must meet certain criteria: reproduction must be suppressed in all females, except in the queen caste; there must be cooperative care of the young; and there must be an overlap between the generations. Naked mole rats are the only mammals to meet all these requirements; and when one studies the challenges they have to face in the desert areas of Kenya, one can appreciate why they have evolved eusociality.

For instance, their food consists of widely dispersed underground tubers. To reach these tubers, they have to construct underground tunnels, a process which requires 3 600 times more energy than is required to walk across the desert surface in search of food. With this prodigious expenditure of energy, they cannot afford too many failures in locating the tubers. Therefore, to maximize the probability of finding a tuber, foraging mole rats will tunnel in different directions, but once a tuber is located, the whole colony shares in the spoils. In contrast, another species of mole rat, which lives in a high rainfall area near Cape Town, has access to a high density of underground tubers and is able to lead a solitary existence. Cooperation in the face of adversity is seemingly not confined to humans.

Continuing our journey south towards the dry riverbed of the Kuiseb, the high dunes on the southern bank of the river begin to take shape. The dunes end abruptly on the southern bank because of the regular flash floods that scour out the river bed. Although the riverbed remains dry for most of the year, there is underground water present in the form of aquifers flowing beneath the surface. The aquifers support the growth of large elegant trees such as *Acacia albida* and the camelthorn. The trees, together with smaller shrubs and herbs, form a riverine community about 200 metres broad, but several hundred kilometres in length. It is in effect a longitudinal oasis. This oasis supports many animal species which cannot be considered typical of the desert because most of these seldom, if ever, wander far from the temperate shade along the riverbed.

We make camp on the northern bank of the river well above the flood line and, as there are still a few hours of daylight left, we clamber down the steep cliff face to the floor of the riverbed. There is a splendid sense of isolation in this wilderness; we are among only a handful of humans present in an area of several thousand square kilometres. Behind a rocky scree is a good place to keep watch over one of the very few natural water holes left in the riverbed since the long drought started. A local baboon troop scoops the sand out of this water hole from time to time, and this benefits a wide spectrum of mammal and bird species.

The first species to attract our attention are the rock martins which belong to the swallow family. They are clicking and twittering as they hawk aerial plankton at speeds of up to 80 kilometres per hour; the white windows in their tails flare as they execute amazingly steep turns next to the cliff face. Namaqua doves, arguably one of the most attractive members of the dove family, drink at the water hole. They are soon joined by red-eyed bulbuls, erecting their cheeky crests and flicking their wings as they jostle at the edge of the water. Several desert bird species, however, can exist entirely without fresh water and never visit the water hole. In these cases sufficient moisture is obtained from their food – usually live arthropods.

Male namaqua sandgrouse, about the size of a large pigeon, are wading into the shallows of the water hole to thoroughly wet their breast feathers. Gordon Maclean of Natal University found that these breast feathers can absorb as much as 40 millilitres of water at a time. The males then ferry the water back to their chicks, in some instances flying over 15 kilometres to deliver their payload. The chicks drink the water directly from the breast feathers; and I have found that they can drink as much as 30 percent of their body weight during this process.

Eventually the baboon troop arrives, much to our excitement. From subtle signs it is clear that they have noticed us but pretend not to do so as they straggle towards the water hole. Conrad Brain and Virginia Mauney have lived for months in this completely isolated canyon, observing this troop. They made the astonishing discovery that the baboons can survive as long as 26 days without drinking water and that, in this dehydrated condition, their body temperatures can rise to as high as 42°C. In humans this would represent a raging fever which would debilitate the patient completely. Towards the end of this period of dehydration they restrict their energy output to a minimum, because activity leads to increased respiration and therefore increased loss of body water from the respiratory system.

They are opportunistic feeders, taking scorpions, fruit, termites, tenebrionid beetles, roots, grass shoots and even the pith of wild tobacco plants. Now, as we watch them, we can see how the water has rejuvenated them. They are far more alert and playful; two males are threatening one another, while a nearby female slaps its infant that has become too boisterous. Suddenly the sentinel gives a warning bark and the troop immediately becomes attentive and moves up the cliff face to avoid an approaching group of large oryx antelope with their long spear-like horns. The baboons will sleep in the cliffs tonight and refresh themselves once again tomorrow morning at the water hole, before disappearing down the canyon in search of scarce and widely distributed food items to satisfy their energy requirements. They are living on the very margin of their natural distribution and definitely walking an energetic tightrope.

Twilight is falling as we climb out of the canyon to watch the sunset on the dune field. The setting sun sharpens the beautifully sculpted ridges and valleys on the dunes; the atmosphere of isolation and unspoilt wilderness brings to mind an anonymous quote, "Of all natural forms, the sand dunes are the most elegant – so simple, severe and bare. Nature in the nude."

As we stroll back to our camp, we can hear a pair of spotted eagle owls performing a duet in the riverine forest. One of our party relates how impressive a full moonrise over the dunes can be, particularly after a strong east wind has been blowing during the day. The huge orange disk appears to float above the dunes and the countless mica dust particles, suspended in the air, sparkle in the strong moonlight. Unfortunately, tonight will be moonless.

Next morning, we waken to discover a dramatic change in our surroundings. The temperature has plummeted and everything is enveloped in a cold, dripping fog that smothers sound like a blanket. We hurry across the river into the dunes to observe how the animals are reacting to this sudden change in the weather. Advective fog off the nearby Atlantic Ocean is a frequent occurrence in the Namib Desert, but mists which condense heavily on the surface of the dunes are rather rare, and we are keen to exploit this opportunity to the full.

As we move up the hard, windward side of a dune, we come across the exquisitely coloured palmato gecko. It is using its webbed feet to dig itself into the side of the dune to escape the fierce heat later in the day. As it digs, its long tongue emerges to lick condensed fog off the side of its head and to clean its eyes of sand grains. It has been hunting small arthropods all night and will spend the entire day underground. Because of its nocturnal habits its skin is mostly without pigment and is so transparent one can see its heart beating. We do, however, wonder why a nocturnal animal should display such a bright turquoise colour on the surface of its head. Is it vestigial coloration from a previous diurnal existence? And how long did it take to evolve the unique characteristic of webbed feet, which facilitates its locomotion and digging in the sand?

We pass a large clump of dune grass, *Stipagrostis sabulicola,* which is dripping with condensed fog. It is almost as if a fine, silent rain is falling. The surface of the sand is also wet and on the steep face of the dune, small pill-shaped beetles have constructed shallow trenches facing into the mist-laden wind. The ridges of these trenches are slightly elevated above the sand surface and trap more condensation than the undisturbed sand surface. The beetles return along these trenches, removing the water from between the sand granules with their specialized mouth parts.

At the crest of the dune, where condensation is at its highest, a solitary large black tenebrionid beetle is fog-basking; facing head down into the gentle fog wind, with its rear end raised into the air. Condensation droplets on the upper surface of the beetle are running down to the mouth parts where they are imbibed. This precious water will be carefully husbanded by the beetle and, in this regard, its almost waterproof cuticle will be a great advantage. Waterproofing in tenebrionid beetles, as with many desert creatures, including reptiles, is due to a thin layer of fatty substances (long chain hydrocarbons) on their outer surface. In addition the beetles have a very low metabolic rate, which reduces water loss from respiration significantly. These observations were made by Mary Seely and Bill Hamilton at the Namib Desert Research Station.

The fog has not destroyed all the evidence of last night's surface activity in the dunes and we are able to identify the tracks of gerbils, beetles, legless lizards and the unique golden mole. This little mammal expends an enormous amount of energy actually swimming through the dune sand on occasion, to hunt for arthropods and, if lucky, legless lizards as well. To compensate for this huge expenditure of energy, the golden mole abandons regulation of its body temperature when not active, and enters a state of torpor, thus saving significant and critical amounts of energy each day.

At the base of the dune there is an unusual plant that has developed into a large thorny thicket. Its species name is *horrida* which probably arose from the fact that the plant is entirely without leaves; it is, instead, covered with sharp green thorns. The thorns and stem of the plant engage in photosynthesis. These nara plants, as they are called by the local Topnaar Hottentots, produce melon-like fruits filled with edible seeds. The seeds, when toasted, taste vaguely like almonds, and archaeological studies have shown that they have been prized as a sweetmeat by humans in the desert for thousands of years.

In the northern Namib a beautiful species of sand lizard, *Angolosaurus skoogi*, relies heavily on the nara for its survival in the coastal dunes. Biologists Duncan Mitchell and Mary Seely have discovered that it feeds on the flowers of nara and also breaks off the tips of the growing points, whereupon a water droplet beads on the broken tip and is imbibed by the lizard. The plants are in a permanent state of positive fluid pressure, or turgor, and obtain their moisture from a very deep tap root that absorbs the moisture accumulating at the base of dune.

The nara plants consolidate the sand around them and gerbils exploit this by constructing their burrows in and around the nara. These attractive little mammals are nocturnal and feed on grass seeds, fresh green material and arthropods. They also eat the carbohydrate-rich seeds of the nara which, after digestion, yield sugars that can be catabolized to carbon dioxide and water. This water is known as metabolic water, or the water of oxidation, and because the gerbil never has direct access to drinking water, it is essential for its survival and has to be carefully conserved. This will be accomplished largely by avoiding the heat of the day in a cool and humid underground burrow. Additionally, like other desert rodents, it conserves large amounts of water by producing a highly concentrated urine. Far less water is therefore lost in ridding the animal of excess salt and other waste products. In the case of Australian hopping mice, the urine concentration reaches the exceptional level of nine times the concentration of sea water. Maximum urine concentration in humans is about the same as the concentration of sea water, which is roughly equivalent to six teaspoons of table salt dissolved in a litre of water.

The dense fog is beginning to thin as it retreats back to the ocean from the dune field. While waiting for the sun to warm the surface of the sand sufficiently to trigger the emergence of the diurnal animals, we reflect on the flow of energy through this unusual dune ecosystem. The dunes, at first glance, appear to be almost vegetationless, but in spite of this, a variety of animals can make a living here. The base of the food chain consists not only of isolated plants which survive on the irregular supply of condensed fog water, but also of wind-blown plant detritus which accumulates in the dunes. This detritus originates from several sources, but a major component originates from the ephemeral growth of desert grasses following a rare rainfall event. This means that the detritus has a high cellulose or fibre content and the next link in the food chain, the tenebrionid beetles, termites and silver fish, must be able to digest the cellulose effectively.

This is usually accomplished by symbiotic micro-organisms living in their digestive tracts.

The next link in the chain consists mostly of spiders, small lizards and a few species of birds such as specialized larks. The very top of the food chain is occupied by the side-winding adder, the golden mole and, on occasion, the greater kestrel and chanting goshawk. In a manner similar to more familiar ecosystems, the available energy declines rapidly as we move up the food chain and, for this reason, the number of animals at the top of the chain are few and there are only three links from the primary producers, the plants, to the top carnivores.

The surface of the sand is now pleasantly warm from the sun and by chance we disturb a large dune spider which, with a little encouragement from the zoologist in our party, is made to sprint to the edge of the dune. On reaching the edge it flips on its side and cartwheels down the face of the dune. We are told that it can reach speeds of 1.5 metres per second using this mode of escape from its major predator, a large black wasp. This is also one of the very few examples of the use of a wheel, or more correctly a rotor, in nature. The discovery of the wheel by humans transformed our lives by greatly reducing the energetic cost of locomotion. Why then have wheels not evolved more commonly in nature? There are several technical reasons for this: for instance, the difficulty of maintaining a blood supply to a freely rotating axle. A more frivolous answer would be that there have not been any roads until recently!

It is heating up rapidly now and the sand surface temperatures are rising towards 50°C. We decide to move back to camp and en route we see several sand-diving *Aporosaura* lizards. They are doing their best to extend their period of activity above the surface of the sand by dancing on the surface, like a child hopping from one foot to another on a hot pavement. It will soon be too hot for them, when they will disappear beneath the sand with a flick of the tail to emerge much later during the late afternoon, when the dune surface cools again. As we return across the river, we catch a glimpse of an adult jackal trotting with minimal energy expenditure in the direction of the water hole.

We break camp and leave in a westerly direction towards the Atlantic Ocean where we intend spending the night on the Skeleton Coast. It is a long, uncomfortable drive, frequently interrupted to examine interesting desert plants, including the remarkable lithops, that resemble the stones dispersed across the soil surface, another example of excellent camouflage in the desert.

We make a special detour into a narrow canyon to view the unique *Welwitschia mirabilis,* which only occurs in the fog belt of the Namib Desert. This plant grows very slowly and close to the ground. It produces only two opposite leaves. These are torn by strong winds into a large number of long narrow leaves, that trail across the ground for several metres. *Welwitschia* is of great scientific interest, because it possesses characteristics of both cone-bearing and flowering plants and is consequently thought to be of ancient origin. The older plants resemble dwarf trees and the male and female cones are borne on separate plants. The botanist in our group tells us that the specimen we are examining, although only 60 centimetres tall, is several hundred years old.

Eventually we reach the ocean, where a cold Southwester is blowing off the Benguela current, which sweeps up the coastline. We shelter behind a tarpaulin tied to ancient whale ribs which, at one time, must have been part of a shelter erected by the stone age beachcombers who lived on this coast. We are grateful for the shelter and the warmth of our driftwood fire; talk soon turns to the desert and the animals we have seen. We get a brief glimpse again of a black-backed jackal sheltering behind a small dune. This is the third jackal we have seen on this brief trip and we agree that they personify the opportunism which is a typical characteristic of desert animals.

We are told that even lions sometimes make their way down the dry river courses from the escarpment to eventually arrive on these desolate desert beaches, where they feed upon dead seals washed up on the shore line – a truly remarkable predator/prey relationship.

Someone then asks if desert animals are really different. The evolutionist in our party says, "Not really," and claims that all desert animals are basically very similar to their close relatives living in temperate ecosystems. They obtain their energy in similar ways and energy flows through a desert ecosystem in the same way as in a temperate system, the only difference being that the desert ecosystem is tightly controlled by sporadic rainfall events. There are (he goes on to say) just fewer animals and they have evolved minor adaptations, mostly behavioural, to fit them better to desert life.

Our physiologist disagrees and protests that the evolution of a mammalian kidney, which can produce urine which is nine times more concentrated than sea water, is not a minor adaptation. And so the debate rages until our ecologist draws our attention to the fact that most animals in the desert are small enough to escape from the heat and desiccating effects of the desert climate, and that they are mostly cold-blooded or, more correctly, ectotherms. The latter characteristic means that their energy requirements are some fifteen times lower than warm-blooded or endothermic creatures, and they can depend upon a reliable source of solar energy to control their body temperatures. Many of them can enter long periods of dormancy, or arrested development (diapause), when energy requirements are reduced to an absolute minimum. In other words, they enjoy all the benefits of a low cost of living. When favourable conditions return, the animals emerge and undergo swift growth and development. This is the reason, she concludes, why the most abundant animal species in the desert are small arthropods and reptiles.

This sounds like sense and we abandon the debate to revel in the solitude and exhilarating environment of this splendid natural laboratory.

Humans
Bend
the
Rules
(From Silicon Tool to Silicon Chip)

If the physicists are correct in explaining the origin of our universe on the basis of the "Big Bang" theory, then *time* did not exist before the Big Bang, which allegedly took place about fifteen billion years ago. Some physicists, however, believe that the universe is much younger. We are also told that the sun's energy will have declined to such an extent five billion years from now that our solar system will no longer exist. It is also a sobering thought that, in spite of the great age of the universe, truly modern humans have trodden the surface of this planet for only 120 000 years, a mere wink of the eye in geological history.

The early existence of modern humans has been described by some archaeologists as brutal, nasty and short; others point out that they may have had a fairly leisurely and carefree existence. We will never know for sure, but it is easy to imagine how many hardships a small family of modern humans (*Homo sapiens*) would encounter in the African savannah, where early humans probably first evolved.

Such a group would most likely be led by the alpha male as they hurry through the tall grass to gain access to a distant shelter. They have been gathering wild figs and the edible roots of various shrubs for most of the day and are exhausted. They relax their almost constant vigil momentarily and are surprised by an adult lioness, whose tawny coat is camouflaged against the blond colour of the tall grass. Without hesitation, the alpha male moves to the front of the group and is just in time to meet the impact of the lion's charge with his raised spear. The spear has been carefully fashioned by fitting a strong wooden shaft to a sharpened thigh bone (femur) of one of the large antelope. The spear point snaps like matchwood and the lion kills the leader swiftly by breaking his neck, but, fortunately for the group, the razor-sharp femur point has severed the carotid artery and jugular vein of the animal on one side of its neck; it is already staggering about from lack of oxygen to the brain. The group naturally notice this and shout and scream encouragement to one another as their adrenalin flows and they pelt the lion with rocks and crude clubs. Eventually the animal succumbs and, using sharp-edged stone tools, the group flay it to gain access to the meat and viscera.

We do not know if the group would have performed a religious ceremony over the dead body of the alpha leader or even have bothered to bury him. Of perhaps more interest is the unhesitating instinct of the leader to risk his life for the group. Evolutionists tell us that this instinctive behaviour contributes to the survival of the leader's close relatives and, therefore, to the partial survival of his genes – a form of selection known as kin selection. This may be another example of "selfish genes" using our bodies to ensure their survival.

The episode with the lioness also serves to illustrate how frequently such a group would experience real fear. Anyone who has encountered dangerous animals in the wild will confirm the surge of adrenalin and the massive stimulation of the flight and fight reflexes by the sympathetic nervous system. This ensures the maximum supply of energy (glucose) and oxygen to the muscles. Such encounters must have been an almost daily occurrence among early humans and so-called "adrenalin-highs" would have been commonplace. This might be a partial explanation of why many young adults, particularly males, still seek thrills today through dangerous sports, such as mountaineering, paragliding and bungee jumping; and vicariously by watching violent films. Even very young children often revel in the telling of the violent aspects of fairy stories. Our physiology cannot have changed greatly in the last forty thousand years and perhaps it is not too far-fetched to believe that we have retained a strong addiction to adrenalin-highs, even though they may leave us weak and trembling after the episode. It is also significant that the widely-abused drugs cocaine and amphetamine produce the equivalent of an adrenalin-high in the brain.

The survival of the family group after their encounter with the lioness would depend on a complex of factors, including escape from predators, avoidance of injury and disease and, above all, the ability to obtain enough energy and nutrients from the environment.

How much energy would these early humans have required and how does this amount compare with other animals? From their skeletal remains they appear to have been smaller than most present-day Europeans and also of a rather slender build. Nevertheless, their active lifestyle of hunting, gathering and scavenging must have demanded considerable energy. The daily energy expenditure of present-day humans has been carefully studied and we have figures for various occupations. For instance, coal miners and farm workers use 3 660 kilocalories per day, university students about 2 930 kilocalories per day and elderly retired people as little as 1 750 kilocalories each day. According to the *New Scientist*, Antarctic explorers Stroud and Fiennes used 11 000 kilocalories per day on their trek across that continent, while Captain Robert Scott's expedition to the south pole (1910-1912) allowed only 4 300 kilocalories per day per person, which presumably led to their tragic deaths from starvation and exposure. If we allow for the smaller size of early humans and assume that their energy expenditure was equivalent to that of a peasant farmer they probably required about 3 000 kilocalories per day, which could rise to well over 4 000 kilocalories when they engaged in long distance running during hunting expeditions.

To satisfy this requirement a Stone Age beachcomber would have to eat about 105 medium-sized marine snails such as *Turbo*. This would have been a reasonably easy task but would not have left a great deal of time over for leisure activities. Alternatively, capturing a small antelope in a snare would allow him to satisfy his requirements with only two large venison steaks of around 500 grams each. Edible roots are difficult to excavate and often contain a great deal of water and indigestible fibre. Consequently, well over 3 kilograms of roots would be required to meet the energy requirements for a single day. Understandably, then, nuts with their high oil content and ripe fruit with a high sugar concentration would have been favourites for reducing the cost of living and bringing refreshing variety to their diet.

Archaeologists are in general agreement that early humans obtained most of their calories from gathering, and that the scavenging of carcasses, killed by the large carnivores, was more important than hunting. Their simple tools allowed them, unlike other scavengers, to gain easy access to the marrow in the long bones of antelope carcasses. Bone marrow is, naturally, a rich source of energy because of its high fat content. The comparatively long legs of humans might also have enabled them to out-run some competing scavengers to the kill; and their naked skin facilitated the cooling effect of sweating while running long distances in a hot environment.

The next question to ask is, How do human energy requirements compare with those of other animals? As we have seen earlier, small flying insects have by far the highest energy requirements and a water-loaded honeybee will use almost one hundred times more energy per unit body weight than an adult human exercising strenuously. Let's, however, be more realistic and confine our comparison to mammals as in Table 1.

Table 1: Resting Metabolic Rate in Selected Mammals

Species	Kilocalories per day	Kilocalories per kilogram body weight per day
Shrew	4	1 000
Rat	29	100
Dog	438	37
Human	1 693	24
Horse	8 179	13
Elephant	31 000	8

The comparisons made in Table 1 show that, although the total kilocalories required daily by present-day humans is more or less what one would predict from the values obtained for other mammals, the human requirements are slightly on the high side. This also holds true when we compare the human metabolic rate per kilogram body weight with the other values. It is, however, important to understand that as the mammals increase in size from that of a shrew to an elephant, their metabolic rates per kilogram body weight decline sharply. This then is the famous elephant to mouse curve, described many years ago by Professor Max Kleiber at the University of California. To date we still do not have a really satisfactory explanation for this phenomenon, but one thing is certain: as animals increase in size, their surface area, relative to their volume, decreases proportionally. Therefore, if an elephant had the same metabolic rate as a shrew, the temperature of its skin surface would have to be well above boiling point, because the surface area of the body is an important determinant of rate of heat loss.

The procurement of energy by social insects and mole rats is greatly facilitated by the division of labour among the individual members of the colony. Similarly, our band of early *Homo sapiens* would have been forced to cooperate closely for survival, assisting each other with the care of infants, finding food, hunting, building of shelters and defense against predators. They would, nevertheless, have been far from eusocial, like the bees, ants and mole rats. Their individual resourcefulness must have been strongly developed, as shown by their subsequent rapid development. Where, however, did this band of *Homo sapiens* originate? Let us now explore how these modest fashioners of silicon tools ultimately developed into the users of silicon chips and enormous amounts of energy.

It is generally accepted today that humans evolved from an ape-like ancestor and that the first hominid species arose well over three million years ago. The fossil remains of these early species, grouped under the genus *Australopithicus*, were first discovered in South and East Africa, and it is very probable that early humans evolved on the African savannahs.

The origin of modern man (*Homo sapiens*), however, remains controversial. Some of the earliest evidence of the existence of modern humans, including a cannibal feast, was found in a cave at the mouth of Klasies River near the southernmost tip of South Africa, near Cape Town. These remains have been dated at 120 000 years ago. Some scientists believe that these modern humans migrated from Africa into Asia and Europe, where they displaced the archaic descendents of *Homo erectus*. This is often referred to as the "Out of Africa Theory." Other scientists, notably in Australia, have suggested that *Homo sapiens* arose from *Homo erectus* in many different regions of the world.

All these species of *Homo* had one important feature in common: an upright stance. We have already discussed the advantages of an upright stance for keeping cool under the warm, dry conditions of the African savannah; but there are other more important advantages, such as being able to observe predators and prey above the tall savannah grasses; to free the hands for using tools and for hurling stones at predators and enemies; and, perhaps most important, to be able to carry food and tools back to a shelter. Our close relative the chimpanzee, which shares more than 98 percent of our genetic material, has difficulty carrying more than three bananas. An upright-stance is, however, not without disadvantages. Locomotion on only two legs is more costly from an energetic point of view than on four legs, and the increased weight that the lumbar vertebrae have to carry can lead to debilitating orthopaedic problems.

Some anthropologists think that, because the upright stance led to a change in the conformation of the human pelvis, the size of the birth canal became reduced and this required that human infants be born earlier in an immature stage of development. This, they believe, subsequently led to greater dependence of the females upon the males while caring for their immature young. This would presumably have stimulated a more social lifestyle within extended family groups. The males would also have had a selfish interest in seeing their offspring survive to maturity, which would have contributed to cementing family relationships.

In any event, the upright stance must have had many advantages as it was adopted by all the species of *Homo* that followed the Australopithicines. These were *Homo habilis* (the handyman), who lived about 2 million years ago at the time the first simple stone tools appeared; then *Homo erectus,* who created advanced tools and lived on Earth for well over a million years; and finally the archaic form of *Homo sapiens,* who appeared some 500 000 years ago. Neanderthal humans emerged some 200 000 years ago and lived in Europe and the Middle East during the last Ice Age. They were also classified as *Homo sapiens,* but disappeared mysteriously about 30 000 years ago. Nevertheless, Western Europeans still retain several Neanderthal features, such as their large noses, and probably many more of which we are still not aware. Our little savannah family, however, belonged to truly modern humans, who have only existed for the past 120 000 years and who, in an act of supreme conceit, have recently named themselves *Homo sapiens sapiens.*

In spite of their recent appearance on our planet, modern humans soon spread to literally all the corners of the Earth, apart from the Antarctic, and soon began to make a significant impact on the environment. Perhaps the best example of the swift expansion of human influence is provided by the invasion of the Americas. There is some doubt as to the exact date when humans first arrived from Siberia in North America across the Bering Strait, when Asia and America were connected by a land bridge. The most generally accepted date is about 12 000 years ago, but new evidence, such as the flaked and fractured bones of mammoths in caves along the Old Crow River in the Yukon, suggest that the date of arrival could be pushed back to at least 24 000 years ago. Whatever the actual date may be, it is remarkable how swiftly these people spread across the Americas to occupy almost every niche available from the frozen North, through prairies, temperate and tropical rain forests to the very tip of South America in Tierra del Fuego. Life within the Arctic Circle must have been the most challenging, demanding a variety of difficult skills for making tools, clothing, shelters and kayaks, as well as fishing and hunting gear. This is perhaps why the impact of the northern people on their environment appears to have been less severe than their cousins who migrated further south.

It seems now that our concept of a pristine environment existing in the Americas prior to the arrival of the Europeans is a myth. Many of the large mammals such as mammoths and sabre-toothed tigers were probably hunted to extinction by humans. William Denevan of Wisconsin University believes that the native Americans had a major impact on shaping the vegetation of North America by using fire, thus promoting the expansion of prairies and park-like ecosystems. Very early farming practices in the highlands of central Mexico caused serious soil erosion, which has not been surpassed since the arrival of the Spanish. Jared Diamond tells of the Chaco pueblos in New Mexico, where the Anasazi cut down over 200 000 juniper trees for roof beams, thereby destroying their forests. Today these dwellings are surrounded by desert.

The increasing impact of humans upon the environment was accompanied by a slow but steady increase in energy use and improvement in their technology. In fact, if I had to suggest a unique characteristic of humankind, I could do little better than to point to the human understanding of lever and pulley mechanics. It is true that several species of animals use tools, and many are excellent architects, but humans are the only primates that *make* tools and, through their understanding of lever mechanics, are able to enhance the energetic efficiency of muscle power. By simply wedging a long pole under a boulder, and using a small stone as a pivot, early humans would have been able to move large boulders with only a moderate expenditure of energy.

The earliest tools discovered to date are those from the Olduvai Gorge in East Africa and date back to about 2.5 million years ago. They are crude choppers made by striking flakes off the end of medium-sized stones. Approximately 400 000 years ago a significant improvement in tool making occurred with the increasing use of bone tools. These tool assemblages are frequently found with the skeletal remains of a variety of ruminant wild animals and other large herbivores, confirming the dependence of early humans upon these animals as a source of nutritional energy. This dependence has persisted throughout most of our history and by the time of the last Ice Age, modern man produced delicate blades and sharp axes in stone by pressure flaking, for butchering mammoths, woolly rhinos and wild horses. The beautiful double-bladed Acheulian hand axe is a classical example of this period. In other words, humans had discovered and exploited the energetic advantage of using sharp edges and points. The small surface area of a sharp edge magnifies the amount of force applied per square centimetre on the edge of the tool, in much the same way as stiletto heels will ruin hardwood floors, and needles will pierce thick leather.

One can also imagine how frequently these early people searched for suitable stony material from which to fashion their tools. This interest in rocks and stones would naturally promote a practical and fairly extensive knowledge of geology and ultimately the discovery of copper ores for producing the first metal tools and weapons. Production of the first metal weapons, however, required the use of fire and although the slow improvement in tool technology contributed significantly to more efficient energy use, probably the most important technological advance in the early history of humankind was the discovery of how to make fire. Bob Brain of the Transvaal Museum has found evidence in the Swartkrans caves near Pretoria, South Africa, that the first use of fire by hominids occurred between 1.0 and 1.5 million years before the present, that is before *Australopithicus robustus* became extinct. He also believes that the ability to make fire represents a watershed in human history, when humans emerged from being a subordinate species to becoming a dominant species.

To digress somewhat, the history of weapons and armaments also mirrors the history of energy use by humans in many ways. The very first weapons would have been made from easily accessible natural materials to make crude wooden clubs and bone daggers. As technology advanced with the discovery of fire-making and the smelting of metals, weapons became more sophisticated. The first example of an organized army going into battle was probably the counter attack of the Egyptians against the Hyksos in 1 600 BC. The Egyptians used chariots and foot soldiers, who were reinforced by archers. The early Greeks used bronze swords very effectively, but these were no match for the short sword (*gladius*) of the Romans which, together with the courage and discipline of the Centurions, transformed the Roman army into a highly efficient killing machine – a necessary adjunct for the expansion of the Roman Empire.

The first weapon used to engage an enemy at a distance beyond arm's length was the slingshot (graphically described in the Old Testament where David slew Goliath). The next important weapon of this nature was the bow and arrow, used extensively by the Assyrian armies some 3 000 years ago.

The development of these light projectiles came as a result of an intuitive appreciation of one of Isaac Newton's famous Laws of Motion, although it was many centuries before Newton was to formulate these laws. Newton's Laws of

Motion have taught us that the kinetic energy (energy of movement) of a moving body is proportional to half the mass of that body multiplied by the square of the speed at which the body is moving. It follows, then, that only a moderate increase in the speed of a projectile will increase its kinetic energy greatly, because speed, or more correctly velocity, is squared to calculate the total kinetic energy. The latter, in turn, largely determines the force with which the projectile will strike the target. This is one of the reasons why the crossbow was so effective, in spite of the arrow being such a light missile. Its velocity compensated for its low mass and it could pierce chain mail at 200 metres. The engineers that designed German artillery for the Great War (1914-18) understood Newton's Laws very well and designed 42 cm (16 inch) guns that could hurl a 1 000 kilogram shell over a distance of 15 kilometres at near supersonic speeds. These cannons were known as Big Berthas and required a crew of over 100 men to operate them.

To return to the historical use of energy by humans, the next major breakthrough was the domestication of wild animals and the beginnings of agriculture, which occurred approximately 10-12 000 years ago. Sheep and goats were domesticated about 10 000 years ago, but it is thought that gazelles were herded under semi-domesticated conditions in the Levant as early as 18 000 B.C. There is reliable archaeological evidence that cattle were domesticated prior to 6 000 B.C.

Pastoral nomadism was a major transition in the lifestyle of humans and soon led to important benefits. First and foremost was the creation of a far more stable source of nutrition. The grazing animals harvested the cellulose in plants and, with the help of the symbiotic micro-organisms in their digestive tracts, split the $\beta1$-4 linkage of the glucose molecules in cellulose to release its constituent sugars. So, in practice, the sheep and goats were harvesting energy for the nomads by producing milk and meat from material that was virtually useless to humans; because, as we have seen, the latter lack the enzymes to break the β1-4 linkage (See Chapter 3).

This stable source of energy and nutrition from grazing animals stimulated many improvements in technology, such as improved shelters, clothing, tools, weapons and cooking utensils. The additional free time was often used for artistic expression in decorated utensils and religious artifacts. One need only reflect on present-day nomads, such as the Bedouins, to understand how efficient nomadism can be. Their camels, sheep and goats use the coarse grasses and shrubs of the desert and provide meat, milk, wool, fat and leather for them. The camels also provide efficient transport, their dung is used as a fuel for cooking and for warmth on cold nights. The wool and leather are fashioned into tents, saddles, harness and clothing. There is time available for fashioning jewellery and other artistic trinkets and utensils. Hunting and gathering still continue, including falconry. No wonder then that this lifestyle, enjoyed within the freedom of endless desert horizons, has been so frequently romanticized.

In more temperate regions, cattle have proved to be particularly valuable as domestic animals, providing meat, milk, leather, butter and the muscular strength to pull ploughs and heavy loads. The latter allowed the transport of food and supplies in bulk and the highly nutritious nature of meat and milk products must have contributed significantly to the growth in population.

Dependence on ruminating animals was not restricted to the Middle East; it soon became widespread throughout the world. It is even reflected today in a popular postcard sold in Cambridge of Kings College in all its splendour. In the foreground there is a broad green pasture with a group of cattle resting and ruminating. Even the "heady" culture of Kings College seems to be based on "chewing the humble cud."

In spite of the great advantage of having a larder on the hoof, so to speak, the nomads could not rely entirely on their herds as a form of stored energy; the wandering lifestyle with temporary dwellings was not conducive to further development. Moreover, eating generously off the second link of the food chain, as opposed to eating the primary producers of energy, the plants, is somewhat of a luxury because of the 90 percent loss of energy that occurs when plant material is turned into animal tissue. Consequently, if humans were to abandon the nomadic and scavenging way of life and settle down in stable communities, they needed a plant product which could be stored for long periods. It also had to provide easily digestible energy, in which the glucose molecules are linked by the α1-4 linkage.

This was eventually obtained from the carbohydrate-rich seeds of two important grass species: barley and wheat. Wild barley had been harvested in the Middle East as long ago as the last Ice Age in Europe, but its first cultivation probably occurred in Syria, Palestine and Mesopotamia about 10-12 000 years ago. It was in many ways the ideal product; the hard seed coat protected the seeds from excessive drying and from becoming spoiled. The carbohydrate constituents were highly digestible and energy-rich after being cooked, and the seeds could be cultivated and stored for long periods with comparative ease. The cooking of hard seeds softened them and reduced the wear on teeth and therefore extended the reproductive lifetime of females, thereby contributing to an increased rate of population growth. The introduction, then, of a reliable energy source in the form of cereal grains was probably the most important revolutionary change in the history of human endeavour since the discovery of fire, almost a million and a half years previously.

In contrast to the long time interval between the discovery of fire and the cultivation of cereal grains, technological advancement subsequent to the introduction of agriculture was very rapid indeed. Within a matter of only 12 000 years humans moved from primitive digging sticks for planting their grain to Saturn rockets for reaching the moon. During this period of rapid change, major civilizations grew and waned. The arts, philosophy and engineering flourished, but it is difficult to select a single example from among these major achievements to illustrate how energy usage by humans had increased simultaneously to keep pace with these cultural developments.

The Egyptian pyramids could be used as a good example, or the era of cathedral building during the thirteenth century, because the latter not only represents a great cultural achievement but also a concentrated expenditure of energy and effort in our history.

The famous Cathedral of Notre Dame at Chartres, often described as one of the greatest buildings ever built, was constructed in 30 short years at the end of the twelfth century. According to Jean Favier, in his absorbing book *The World of Chartres*, the cathedral was built by nine teams of masons, each under the control of a master mason. These teams may have consisted of as many as 300 men (let us assume an average of 200). Incidentally, the teams resided in wooden buildings known as lodges; and some historians believe these associations eventually gave rise to the Freemason movement.

If we allow the established daily calorie requirement of 4 000 kilocalories for stone masons and a daily work force of 1 800 men (9 x 200) working on average 300 days per year; then the annual cost of construction can be estimated at 2 160 000 000 kilocalories, or 64 800 000 000 kilocalories for the total period of 30 years. Then, if we apply our universal currency of kilograms of wheat to overcome inflation, the total cost of the cathedral was approximately 19 058 824 kg of wheat. Bearing in mind that many other splendid Romanesque and Gothic cathedrals were under construction at the same time, this was a remarkable achievement for the non-mechanized agriculture of that period; and the European cathedrals must be considered fitting monuments to the spirit of the men and women of those times.

About 400 years after the end of the intensive period of cathedral building, a most influential scholar in the person of Isaac Newton appeared on the scene. His contributions towards mathematics, mechanics and theories of gravitation were profound and formed the basis of classical mechanics. This new knowledge resulted in the development of sound engineering principles to replace the guess work, recipes and "trial and error" approach of former years. Needless to say, these developments also led to even greater exploitation of our energy resources, culminating some two hundred years later, when we learnt to harness a super-abundance of energy in the form of fossil fuels. Coal soon began to play a key role in industry by driving steam engines, by providing coke, coal gas, coal tar and eventually the production of electricity, fertilizers and plastics. Modern western society would collapse almost overnight if oil were removed from our economy.

This new period also requires a symbol to exemplify its urgency and excitement. We could perhaps point to the influence of Albert Einstein and other famous physicists of the 20th century, whose work ultimately led to the transformation of a small amount of matter into an enormous amount of energy during the first nuclear explosion in New Mexico on July 16, 1945. We could highlight the achievements of molecular genetics and the recent first cloning of a mammal in Scotland. Alternatively, we could cite the profound influence that computers are having on our daily lives. Communications via the internet are reaching the most remote villages of the world at almost the speed of light and, with just the touch of a key, we will soon be able to gain access to the holdings of the Library of Congress. We have indeed come a long way from the first crude silicon (stone) tools to the silicon chip that controls our computers. These examples might, however, be too obvious and because there are very few new cathedrals to point to, perhaps space exploration and the Saturn rockets would be the most appropriate symbol of our ultimate engineering endeavour of the 20th Century.

If, for example, we again use kilograms of wheat as an understandable common energy currency to overcome the complication of inflation, then the cost of the Apollo programme to land the first man on the moon was equal to about 671 000 000 000 kilograms of wheat. This figure is based on the United States space budget between 1962 and 1969 and the 1969 wholesale price of wheat. Table 2 shows then that the cost of living for Stone Age people and dinosaurs, and the cost of building mediaeval cathedrals is far more modest than space exploration. Moreover, when we remember that the Apollo programme represents a relatively minor item when compared with the total consumption of energy for defense and industry in the world today, then we must agree that the age of super-extravagance in energy use has indeed arrived.

Table 2: Comparing Energy Costs

Stone Age Man (45 kg)	= 1 kg wheat / day
Dinosaur (50 000 kg)	= 76 kg wheat / day
Chartres Cathedral (Total cost)	= 19 058 824 kg wheat
Apollo 11 (Total cost)	= 671 000 000 000 kg wheat

To gain a more down-to earth insight into the dramatic increase in energy consumption, let's first return to our little family band of *Homo sapiens sapiens* in the African savannah some 40 000 years ago.

They have had a successful but hard day hunting antelope, which they were able to drive into a small box canyon before killing three animals with their bone-tipped spears. They are now resting on a high rocky ridge, overlooking an extensive bushveld or savannah plain. Their backs are turned to a temporary shelter made of thorny Acacia trees, felled with stone axes. Evening is approaching and, apart from a young male knapping sharp stone points for tomorrow's hunt, most people are resting. A large fire is burning in anticipation of the barbecue. It is tended by young mothers with very small infants, who remained in the encampment during the hunt. Although there is a sharp distinction in the division of labour between the sexes, there is a general sense of co-operation, but at the same time, a spirit of resourcefulness and individual enterprise is apparent as well. The concept of unemployment does not exist.

We do not know much about communication among these people, although their brain size suggests that they had long since developed speech. We can, however, deduce that they must have been excellent ecologists. They would have known a great deal about edible plants, toxins, the anatomy and behaviour of animals as well as the effect of seasons upon plant and animal life. Injuries must have been comon and the control of pain amajor challenge. Because of their low population density and their simple requirements for food, animal skins and stone tools, their impact on the environment could not have been too great; apart perhaps from the use of fire to improve and expand the grazing for several important species of antelope.

In complete contrast, we can look in on a suburban family of four in Toronto. Their spacious three-bedroom house, with two bathrooms, is centrally heated and cooled to provide a year-round indoor temperature of 20°C. Each member of the family has an automobile, except for the elder son who is unemployed. The parents prefer to drive powerful American cars with 4.5 litre engines. Both have remote controls to switch their cars' heaters on for 10 minutes before they leave their offices in winter. At home the cars remain overnight in a heated garage. They are all busy people and do not have time to cook, other than to heat up pre-cooked meals in a microwave oven. Their clothes are washed and dried in automatic machines and the large garden is sprayed, fertilized and mown by a gardening service for a fee. There are three television sets, a personal computer and a video recorder in daily use.

If we should now try to calculate the energy consumption of this family, we would not only have to include their direct use of natural gas, gasoline and electricity but the indirect services they enjoy as well. All the people who have manufactured, sold, delivered and maintained the many items they use daily also have to eat, heat their houses, buy clothes and possibly drive automobiles. This makes it a complex calculation, but even a superficial reckoning tells us that the energy consumption of our Toronto family must be at least an order of magnitude greater than our Stone-Age family, and that their impact on the environment is cause for grave concern.

Although this extravagant use of energy has led to a great increase in prosperity in the West, many serious environmental problems have arisen which are well known and hardly need elaboration. The excessive burning of fossil fuels (about 70 million barrels of oil daily) has polluted the air, causing acid rain, and now threatens the world's climatic stability through the greenhouse effect. Population growth is still out of control, while natural ecosystems are being plundered as biodiversity declines precipitously. Industrial effluent, toxic waste, insecticides, fertilizers and herbicides are poisoning our rivers, lakes and oceans. Violent crime has become an acceptable risk in most of our major cities, and barbaric civil wars involving torture and mutilation are still, at the time of writing, being fought in Eastern Europe and Africa. Perhaps even more disturbing is that effective government is disappearing from many countries, and the amount of productive land and available water per capita is declining alarmingly, as the world's population continues to increase towards 8.5 billion in the year 2025. According to David Suzuki, "it took 2.5 million years for our species to reach a population of one billion. Today we are adding a billion people every 11 years."

What, then, of the future? Is there any hope of reversing these trends? The cynics will say "No" and there is no shortage of them; but some distinguished scientists, like Jared Diamond of the University of California, remain cautiously optimistic, reminding us that our environmental problems are our own doing and that we should, therefore, be able to solve these problems with our many unique attributes.

In this regard we should accept that although modern technology has played an important role in destroying our natural environment, it is vital to understand that it is not the fault of technology as such, but the way it has been used.

In fact, our future efforts to protect and clean up the environment will depend heavily on the use of modern, innovative technology. What is needed, however, is a change in values – values that may be in conflict with the short-term demands of our selfish genes, but ones that cherish all forms of life and reject greed and violence.

No one, of course, can predict what the short and medium-term future holds for us with any certainty. In the long-term, however, we can be certain that long before the next five billion years have passed, we shall have run out of our primary source of energy. Our sun will probably have imploded into a black hole and life will have ceased to exist on this planet. Professor Boothroyd of the University of Toronto believes we will have to evacuate the planet much sooner than this; he has predicted that the sun will brighten by as much as 10% within 1.1 billion years, destroying all life in the process.

This leaves us enough time to plan our escape strategy to another solar system. Nonetheless, I cannot overcome the anxiety that, because our genes have used our bodies as mere instruments for their survival and reproduction for millions of years, there will only be room on the space ship for our "selfish genes."

This will naturally be their ultimate triumph. But let's not be small-minded and rather wish them *Bon Voyage!*

Suggested Further Reading

Ackerman, D. (1995) *A Natural History of the Senses*. Vintage, New York.

Alexander, R.M. (1982) *Locomotion of Animals*. Blackie, London.

Andrews, M. (1991) *The Birth of Europe*. BBC Books, London.

Asimov, I. (1962) *Life and Energy*. Doubleday, New York.

Clutton-Brock, T.H., Guinness, F.E. and Albon, S.D. (1982) *Red Deer: Behavior and Ecology of Two Sexes*. The University of Chicago Press, Chicago.

Colinvaux, P. (1978) *Why Big Fierce Animals are Rare*. Allen and Unwin, Boston.

Darling, F.F. (1956) *A Herd of Red Deer*. Oxford University Press, Oxford.

Dawkins, R. (1976) *The Selfish Gene*. Oxford University Press, Oxford.

Desmond, A.J. (1975) *The Hot-Blooded Dinosaurs*. Blond and Briggs, London.

Dethier, V.G. (1962) *To Know a Fly*. McGraw-Hill, New York.

Diamond, J. (1992) *The Third Chimpanzee*. Harper Collins, New York.

Downes, S. (1983) *The New Compleat Angler*. Stackpole Books, Harrisburg, Pennsylvania.

Favier, J. (1990) *The World of Chartres*. H.N. Abrams, New York.

Forsyth, A. (1986) *A Natural History of Sex*. Chapters, Shelburne, Vermont.

Goody, J. (1982) *Cooking, Cuisine and Class*. Cambridge University Press, Cambridge.

Gould, S.J. (1995) *Dinosaur in a Haystack: Reflections in Natural History*. Harmony Books, New York.

Heinrich, B. (1990) *Ravens in Winter*. Vintage, London.

Hofmann, R.R. (1989) Evolutionary steps of ecophysiological adaptation and diversification of ruminants: a comparative view of their digestive systems. *Occologia 78*: 443-457.

Louw, G.N. (1993) *Physiological Animal Ecology*. Longman, London.

Logsdon, J.M. (1970) *The Decision to Go to the Moon*. M.I.T. Press, Cambridge, Massachusetts.

McGee, H. (1984) *On Food and Cooking: The Science and Lore of the Kitchen*. MacMillan, New York.

Mead, C. (1983) *Bird Migration*. Country Life Books, Feltham, Middlesex.

O'Connell, R.L. (1989) *Of Arms and Men*. Oxford University Press, New York.

Papi, F. (editor) (1992) *Animal Homing*. Chapman and Hall, London.

Ponting, C. (1991) *A Green History of the World*. Sinclair-Stevenson, London.

Putman, R. (1988) *The Natural History of Deer*. Christopher Helm, London.

Ridley, M. (1985) *The Problems of Evolution*. Oxford University Press, Oxford.

Rosenzweig, M.L. (1974) *And Replenish the Earth*. Harper and Row, New York.

Schmidt-Nielsen, K. (1972) *How Animals Work*. Cambridge University Press, London.

Schrödinger, E. (1946) *What is Life?* Cambridge University Press, Cambridge.

Seely, M.K. (1992) *The Namib*. John Meinart (Pty) Ltd., Windhoek, Namibia.

Smith, J.M. (1978) *The Evolution of Sex*. Cambridge University Press, Cambridge.

Smith, H.W. (1953) *From Fish to Philosopher*. Little, Brown and Co., Boston.

Tattersall, I. (1993) *The Human Odyssey*. Prentice Hall, New York.

Wehner, R. (1987) "Matched Filters" – neural models of the external world. *Journal of Comparative Physiology A.* 161, 511-31.

Wilson, E.O. (1992) *The Diversity of Life*. Allen Lane, The Penguin Press, London.

Glossary

adrenaline: a hormone produced by the adrenal gland; initiates fight and flight reflexes.

alkaloids: a group of nitrogenous plant products such as nicotine and opium; often poisonous.

allergenic: capable of inducing an allergy.

amino acids: organic acids containing an amino group; they form the building blocks of proteins.

arterial: pertaining to the blood system which conveys oxygenated blood from the heart to the rest of the body.

asexual reproduction: reproduction without fusion of male and female gametes.

atom: the smallest part of an element that can take part in a chemical reaction.

ATP: adenosine triphosphate; the major energy-carrying molecule in the cell.

azimuth: the angular distance measured from the south point of the horizon in astronomy.

biome: a major regional ecological community, characterized by distinctive life forms, e.g., boreal forest.

blubber: a thick layer of fat beneath the skin of aquatic mammals.

brown adipose tissue (BAT): specialized fat tissue, capable of producing large amounts of heat while wasting ATP; especially important in hibernators and infants.

caecum: a blind pouch off the large intestine.

carbohydrate: large group of organic compounds, which include sugars, starch and cellulose.

carotid rete: a fine network of blood vessels at the base of the brain, which serves to cool hot arterial blood before it enters the brain.

catecholamines: a group of compounds which include the adrenalines.

cellulase: a rare enzyme capable of splitting cellulose into its constituent sugars.

cellulose: a large molecule consisting of chains of glucose molecules joined by the β-linkage; major constituent of cell walls of green plants.

corpus luteum: yellow body formed from the ovarian follicle; secretes progesterone.

cortisol: a major hormone of the adrenal cortex.

crop milk: nutrient solution secreted by the wall of the crop in certain birds such as the pigeon; regurgitated to feed their chicks.

DNA: deoxyribonucleic acid: consists of two complimentary nucleotide chains arranged in a double helix; carries the code of genetic information for cells.

ectopic: displaced to an abnormal position.

ectotherm: animal that regulates its body temperature by using an external heat source – usually the sun.

electron: negatively charged subatomic particle; orbiting the atom's positively charged nucleus.

endotherm: an animal that produces heat internally to regulate its body temperature.

entropy: the degree of disorder within a system.

esters: organic compounds produced by reaction between acids and alcohols; often have a fruity flavour.

exoskeleton: tough outer covering of insects and other arthropods, which acts as a skeleton.

fatty acids: organic acids consisting of long hydrocarbon chains, which can be either saturated or partially saturated with hydrogen atoms; major constituent of fats and oils.

flight energetics: the energetic cost of flying.

forbs: broad-leafed herbaceous plants.

fossil water: water that has been impounded underground for thousands of years.

free water: water that is not chemically bonded to another compound.

gall: abnormal outgrowth of tissue on the surface of a plant; usually induced by pathogens or parasites.

gene: a sequence of nucleotides in a DNA molecule which functions as unit of heredity.

gestation: pregnancy.

guanine: a nitrogen-rich compound found in urine and tissues of certain animals.

heat capacity: amount of heat required to warm 1 gram of a substance by 1°C.

herbicide: chemical used to kill plants.

hyperthermia: body temperature above normal.

hypothalamus: a small part of the base of the brain; mainly involved in controlling thermoregulation, reproduction, appetite, water balance and the function of the pituitary gland.

hypothermia: body temperature below normal.

indoles: putrid-smelling organic compounds mostly produced in the lower digestive tract.

infrasound: low frequency sound, below threshold of human hearing.

isotope: element differing in mass or radio-activity from an element occupying the same place in the periodic table. Frequently used as tracers in biological research.

kinetic energy: energy of motion.

lumbar: pertaining to the lower back.

metabolic rate: the rate at which an organism uses energy.

methane: marsh gas; denoted by the formula CH_4.

mitochondrion: a subcellular structure in which the Krebs cycle takes place and where most ATP is produced.

Molar Solution: solution of water that contains 1 gram molecular weight of the dissolved substance in 1 litre of the solution.

molecule: a particle consisting of two or more atoms held together by a chemical bond.

myrmecologist: a scientist who studies ants.

non-shivering thermogenesis (NST): physiological production of body heat by means other than shivering – usually by rapid breakdown of fat in BAT.

oestrogen: female sex hormone, mainly produced by the ovarian follicle.

olfactory: pertaining to the sense of smell.

opiate: compounds having similar properties to opium.

oxytocin: hormone produced in the hypothalamus and secreted by the pituitary gland. It causes smooth muscle to contract; e.g. contraction of the uterus during the birthing process.

pellagra: a disease caused by a deficiency of the vitamin niacin; characterized by skin reddening and scaling.

pH: hydrogen ion concentration of a solution; a measure of the degree of acidity and alkalinity.

photoperiod: the rate of change of daylight length.

pituitary gland: small bilobed endocrine gland situated just below the base of the brain. It secretes a number of vitally important hormones controlling metabolism and reproduction.

progesterone: reproductive hormone produced by the corpus luteum in the ovary; responsible, among other functions, for the maintenance of pregnancy.

proton: a subatomic particle with a single positive charge equal to the charge of an electron.

pueblo: communal village of certain Native Americans.

retina: photosensitive portion of the eye.

rumen: capacious forestomach of ruminating animals.

rut: mating season.

savannah: open grasslands with scattered trees and/or shrubs.

silicon: abundant element occurring principally in sand and rock formations.

tannins: astringent plant products used as a defense against herbivores.

testosterone: male sex hormone secreted by testes of vertebrate animals.

thermal inertia: degree of resistance to heating and cooling; largely determined by body size.

thermal panting: panting to enhance evaporative cooling of the body; in contrast to panting to repay oxygen debt after heavy exercise.

thermodynamics: the study of the transformation of energy.

thermoregulation: regulation of body temperature in animals.

torpor: a state of short term dormancy as opposed to long term hibernation.

tracheoles: fine system of tubules which convey oxygen directly to insect tissues.

turgor: the rigidity of plant cells and organs caused by hydrostatic pressure exerted on cell walls.

vector: compass direction.

venous: pertaining to the blood system that returns blood to the heart from peripheral tissues.

wing loading: the mass of a flying animal divided by the surface area of the wings.

ENERGY UNITS
1 calorie = 4.184 joules

1 kilocalorie = 1 000 calories

1 watt = 1 joule per second